Thermodynamics of Surfaces and Interfaces

An accessible yet rigorous discussion of the thermodynamics of surfaces and interfaces, bridging the gap between textbooks and advanced literature by delivering a comprehensive guide without an overwhelming amount of mathematics.

The book begins with a review of the relevant aspects of the thermodynamics of bulk systems, followed by a description of the thermodynamic variables for surfaces and interfaces. Important surface phenomena are detailed, including wetting, crystalline systems (including grain boundaries), interfaces between different phases, curved interfaces (capillarity), adsorption phenomena, and adhesion of surface layers. The later chapters also feature case studies to illustrate real-world applications. Each chapter includes a set of study problems to reinforce the reader's understanding of important concepts, with solutions available for instructors online via www.cambridge.org/meier.

Ideal as an auxiliary text for students and as a self-study guide for industry practitioners and academic researchers working across a broad range of materials.

Gerald H. Meier is the William Kepler Whiteford Professor of Materials Science at the University of Pittsburgh. He has authored or co-authored two books and 175 articles, and has worked as a research collaborator or consultant with many companies in the gas turbine and aerospace industries. He has been a Fellow of ASM International since 1996.

Cover Description

A collection of EBSD images (brightly colored) from several steels and SEM micrographs of the same areas overlayed with misorientation points to describe the misorientation across the grain boundaries.

The author gratefully acknowledges Ms. Rita Patel and Dr. Raymundo Ordonez of the University of Pittsburgh for providing these images.

Thermodynamics of Surfaces and Interfaces

Concepts in Inorganic Materials

Gerald H. Meier
University of Pittsburgh

CAMBRIDGE
UNIVERSITY PRESS

University Printing House, Cambridge CB2 8BS, United Kingdom

Cambridge University Press is part of the University of Cambridge.

It furthers the University's mission by disseminating knowledge in the pursuit of education, learning and research at the highest international levels of excellence.

www.cambridge.org
Information on this title: www.cambridge.org/9780521879088

© Gerald H. Meier 2014

This publication is in copyright. Subject to statutory exception
and to the provisions of relevant collective licensing agreements,
no reproduction of any part may take place without the written
permission of Cambridge University Press.

First published 2014

Printed in the United Kingdom by TJ International Ltd. Padstow Cornwall

A catalog record for this publication is available from the British Library

Library of Congress Cataloging in Publication data
Meier, Gerald H., 1942– author.
Thermodynamics of surfaces and interfaces : concepts in inorganic materials / Gerald H. Meier, University of Pittsburgh.
    pages cm
Includes bibliographical references and index.
ISBN 978-0-521-87908-8 (hardback)
1. Surfaces (Physics)   2. Surface chemistry.   3. Interfaces (Physical sciences)
4. Thermodynamics.   I. Title.
QC173.4.S94M45   2014
530.4′17 – dc23   2013049905

ISBN 978-0-521-87908-8 Hardback

Additional resources for this publication at www.cambridge.org/9780521879088

Cambridge University Press has no responsibility for the persistence or accuracy of URLs for external or third-party internet websites referred to in this publication, and does not guarantee that any content on such websites is, or will remain, accurate or appropriate.

This book is dedicated to my loving wife, JoAnn, with much affection and appreciation.

Contents

Preface	*page* xiii
Acknowledgements	xv

1 Summary of basic thermodynamic concepts 1
  1.1 Basic thermodynamics 1
    1.1.1 Extensive and molar properties of a thermodynamic system 1
    1.1.2 The first law 3
    1.1.3 The second law 5
    1.1.4 The third law 6
    1.1.5 Combined first and second laws 7
  1.2 Multicomponent systems – solution thermodynamics 10
    1.2.1 The ideal-solution model 12
    1.2.2 Non-ideal solutions 12
  1.3 Multiphase equilibria 17
    1.3.1 Unary systems 18
    1.3.2 Multicomponent systems 21
  1.4 Chemical reactions 30
    1.4.1 Chemical reactions involving gases 32
  1.5 Summary 34
  1.6 References 34
  1.7 Study problems 35
  1.8 Selected thermodynamic data references 38

2 Introduction to surface quantities 40
  2.1 Description of a surface/interface 40

	2.2	Thermodynamic properties	43
		2.2.1 Creation of a surface	45
		2.2.2 Extension of a surface	45
		2.2.3 Relations among surface quantities	47
		2.2.4 Relations between γ and σ	50
		2.2.5 Determination of surface parameters	54
		2.2.6 Description of surface contributions to the thermodynamic description of material systems	67
	2.3	Summary	69
	2.4	References	70
	2.5	Study problems	71
3	Equilibrium at intersections of surfaces: wetting		73
	3.1	Non-reactive versus reactive wetting	73
	3.2	Non-reactive wetting	74
		3.2.1 The contact angle on an ideal solid surface (Young's equation)	74
		3.2.2 Work of adhesion	77
		3.2.3 Capillary rise	78
		3.2.4 Small droplets	80
		3.2.5 Non-ideal surfaces	80
	3.3	Reactive wetting	87
	3.4	Selected values of interfacial energies	91
	3.5	Summary	91
	3.6	References	91
	3.7	Study problems	92
4	Surfaces of crystalline solids		94
	4.1	Surface energy for crystalline solids	94
		4.1.1 Equilibrium crystal shape	98
	4.2	Internal boundaries	102
		4.2.1 Types of grain boundaries	102
		4.2.2 Intersections of grain boundaries with free surfaces	113

	4.2.3 Intersections of grain boundaries	115
4.3	Faceting	116
4.4	Measurement of surface and grain-boundary energies	120
	4.4.1 The zero-creep technique	120
	4.4.2 The multiphase-equilibrium (MPE) technique	123
	4.4.3 Selected values of high-angle grain-boundary energies	124
4.5	Summary	124
4.6	References	125
4.7	Study problems	126
5	**Interphase interfaces**	**128**
5.1	Interface classifications	128
	5.1.1 Coherent interfaces	128
	5.1.2 Semicoherent interfaces	132
	5.1.3 Incoherent interfaces	133
	5.1.4 Interface mobility	133
5.2	Interaction of second phases with grain boundaries	134
5.3	Thin-film formation	135
	5.3.1 Growth of thin oxide films	137
	5.3.2 Formation of metal films by evaporation	143
5.4	Summary	145
5.5	References	146
5.6	Study problems	147
6	**Curved surfaces**	**148**
6.1	Derivation of the Laplace equation	148
	6.1.1 Techniques that use the Laplace equation to measure surface energy	151
6.2	The effect of curvature on the chemical potential	153
	6.2.1 Grain growth	156
6.3	Phase equilibria in one-component systems	158
	6.3.1 The relation between μ_S and μ_L (or μ_V)	158

	6.3.2 The vapor pressure of a pure liquid	160
	6.3.3 The vapor pressure of an isotropic solid particle	162
	6.3.4 The melting point of a one-component solid	164
6.4	Nucleation	165
	6.4.1 Homogeneous nucleation	166
	6.4.2 Heterogeneous nucleation	168
6.5	Phase equilibria in multicomponent systems	168
	6.5.1 The vapor pressure of a component over a multicomponent liquid	168
	6.5.2 The effect of particle size on solubility	170
	6.5.3 Precipitate coarsening	176
6.6	Summary	179
6.7	References	180
6.8	Study problems	181

7 Adsorption 184
  7.1 The Gibbs adsorption equation 186
      7.1.1 Applications of the Gibbs adsorption equation 188
  7.2 The Langmuir adsorption equation 191
  7.3 The effects of adsorption on the fracture of solids 195
      7.3.1 The effect of water vapor on the fracture of ceramics 195
      7.3.2 The effect of grain-boundary segregation on the fracture of metals 198
  7.4 Summary 207
  7.5 References 207
  7.6 Study problem 209

8 Adhesion 210
  8.1 The origin of stresses in multilayer systems 211
      8.1.1 Formation stresses 211
      8.1.2 Thermal stresses 212
      8.1.3 Applied stress 214

8.2 Response to stress   215
    *8.2.1 The relation of the fracture energy and the
            work of adhesion*   218
    *8.2.2 The effect of adsorption on the work of
            adhesion and fracture energy.*   220
8.3 Case study – protective layers on superalloys in
     gas turbines   221
    *8.3.1 Formation and adhesion of protective oxide
            layers*   221
    8.3.2 Multilayer systems – thermal barrier coatings   226
8.4 Summary   233
8.5 References   234
8.6 Study problems   235

Index   237

Preface

There are two objectives of writing this book. Firstly, the subject of thermodynamics, as it is usually taught in undergraduate courses in Materials Science, Chemistry, Chemical Engineering and Mechanical Engineering, does not include a treatment of surfaces and interfaces, or includes only a cursory treatment. The major reason for this is the lack of a suitable text. Some books do not include the subject at all and others contain only a single chapter. The treatment in the latter is often very condensed. On the other hand, there are excellent monographs on the subject, but these are too large, intimidating and/or expensive for use in undergraduate and lower-level graduate courses. The purpose of this book is to bridge the gap by providing a text that is complete and rigorous enough to be the basis for an auxiliary section in a basic second thermodynamics course. Alternatively, it may be the primary text for a course dedicated to the thermodynamics of surfaces and interfaces in which the instructor would present case studies in addition to those already in the text.

Secondly, there are many young and middle-aged professionals whose formal education lacked a substantial treatment of the thermodynamics of surfaces and interfaces for the reasons described above. Nevertheless, an understanding of the subject is important in their day-to-day activities. These include professionals working in aqueous and high-temperature corrosion, coatings, microelectronics, welding and brazing and various applications of nanostructures. This book provides a straightforward discussion of the thermodynamics of surfaces and interfaces that such professionals should find useful.

In order to keep to the two objectives, there has been no attempt to provide an exhaustive review of the literature. This would increase the factual content without necessarily improving the reader's understanding of the subject and would, therefore, increase both the size and the price of the book without enhancing its usefulness as an introduction to the subject. Extensive literature quotation is already available in books previously published on the subject and in review articles. Similarly, the treatment of techniques of investigation of surfaces and interfaces has been restricted to a level that is sufficient for the reader to understand how the subject is studied without involving an overabundance of experimental details. Such details are available elsewhere, as indicated. The prime intent is to provide a background for reading the literature and further independent study.

The book begins with a review of the relevant aspects of the thermodynamics of bulk systems (Chapter 1). It then includes a description of the thermodynamic variables employed to describe the behavior of surfaces and interfaces (Chapter 2). In this chapter the distinction between *surface energy* and *surface stress* is made. Then important surface phenomena are described. These include wetting (Chapter 3), surfaces and interfaces in crystalline systems, including grain boundaries (Chapter 4), interfaces between different phases (Chapter 5), curved interfaces (Chapter 6), adsorption phenomena (Chapter 7) and adhesion of surface layers (Chapter 8). The later chapters also contain case studies to illustrate the application of the concepts that are developed. Each of the chapters contains a set of study problems to reinforce the reader's understanding of important concepts.

Acknowledgements

Dr. M. A. Helminiak and Dr. N. M. Yanar of the University of Pittsburgh are thanked for their assistance in preparing figures for this book. Dr. M.-J. Hua, also of the University of Pittsburgh, kindly provided several high-quality micrographs for inclusion in the book. Ms. Eileen Burke is acknowledged for assistance in preparing the manuscript.

The author also greatly appreciates helpful discussions with Professors A. J. DeArdo and G. Wang of the University of Pittsburgh.

The author acknowledges the many discussions on thermodynamics over the years with his colleague W. A. Soffa, Professor Emeritus of the University of Virginia, which inspired the preparation of this text.

Finally, the author gratefully acknowledges the patience and support of his wife, JoAnn, to whom this book is dedicated.

1 Summary of basic thermodynamic concepts

This chapter provides a summary of the three laws of thermodynamics and the important defined functions and relations for applying these laws to materials systems. It is assumed that the reader has completed an introductory course on thermodynamics. The purpose of this chapter is to bring the reader back "up to speed". An extensive reference list of thermodynamic data sources is also provided.

1.1 Basic thermodynamics

The subject of thermodynamics is based on three empirical laws and their application, generally through the use of specially defined functions. A summary of the three laws and the various defined functions follows. The reader is referred to one of the many comprehensive texts on thermodynamics for a more detailed treatment [1–4].

1.1.1 Extensive and molar properties of a thermodynamic system

The properties (state functions) which refer to the entire system and, therefore, are dependent on size (e.g. mass, volume) are termed *extensive* and may be represented by a generic quantity, Q'. Those properties which are independent of the size of the system (e.g. temperature, pressure) are termed *intensive*. The ratio of any two extensive properties becomes an intensive property. A particularly useful quantity of this type arises when a particular Q' is divided by the number of moles of material in the system, yielding a *molar quantity*, Q:

$$Q = \frac{Q'}{n} \tag{1.1}$$

For example, $V = V'/n$ is the molar volume of the system.

The contribution of each component to an extensive property of the system under isobaric and isothermal conditions is described by the *partial molar quantities*, $\bar{Q}_i$:

$$\bar{Q}_i \equiv \left(\frac{\partial Q'}{\partial n_i}\right)_{T,P,n_j} \tag{1.2}$$

where n_i represents the number of moles of component i and n_j represents the numbers of moles of the other components in the system. $\bar{Q}_i$ is that part of Q' which is contributed by one mole of component i. This is expressed as follows:

$$Q' = \sum_i n_i \bar{Q}_i \tag{1.3}$$

The important *Gibbs–Duhem* relation between the partial molar quantities [5] is obtained by combination of the definition of Q' in differential form,

$$dQ' = \sum_i n_i \, d\bar{Q}_i + \sum_i \bar{Q}_i \, dn_i \tag{1.4}$$

with the mathematical properties of $Q'(T, P, n_1, n_2, \ldots)$

$$dQ' = \left(\frac{\partial Q'}{\partial T}\right)_{P,n_i} dT + \left(\frac{\partial Q'}{\partial P}\right)_{T,n_i} dP + \sum_i \bar{Q}_i \, dn_i \tag{1.5}$$

to yield

$$-\left(\frac{\partial Q'}{\partial T}\right)_{P,n_i} dT - \left(\frac{\partial Q'}{\partial P}\right)_{T,n_i} dP + \sum_i n_i \, d\bar{Q}_i = 0 \tag{1.6}$$

Similarly, the molar and partial molar quantities may be related and for the simple case of a binary system A–B with the mole fraction of

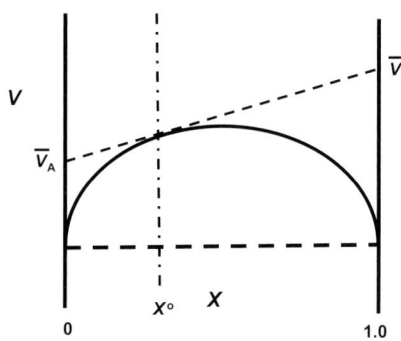

Figure 1.1 Plot of molar volume versus mole fraction for a hypothetical binary system A–B. The intercepts of the tangent drawn at X° are the partial molar volumes.

component B represented by X the relations are

$$\bar{Q}_A = Q - X \frac{dQ}{dX} \tag{1.7}$$

$$\bar{Q}_B = Q + (1 - X) \frac{dQ}{dX} \tag{1.8}$$

These relations [5] have the simple graphical interpretation that the intercepts of a tangent to a plot of Q versus X at $X = 0$ and $X = 1$, respectively, are $\bar{Q}_A$ and $\bar{Q}_B$. The use of these relationships is illustrated in Figure 1.1 for a hypothetical binary system A–B. In this case Q represents the molar volume of the system, V. The tangent line drawn at composition X^0 has intercepts $\bar{V}_A$ at $X = 0$ and $\bar{V}_B$ at $X = 1.0$, which are the partial molar volumes of A and B, respectively, for a solution of that particular composition.

1.1.2 The first law

The *first law of thermodynamics* is a formulation of the *law of conservation of energy*. For a closed system without chemical reaction it may be written as

$$dE' = \delta q + \delta w \tag{1.9}$$

where E' is the extensive *internal energy* of the system and is a state function, q is the *heat absorbed by the system* and w is the *work done on the system*. The heat and work are non-state functions (i.e. they are path-dependent), hence the symbol δ is used, rather than the usual d, for their differentials.

1.1.2.1 Work

Work can be done on the system by a variety of forces but can always be written according to the mechanics definition of the force multiplied by the distance through which it acts,

$$\delta w_i = F_i \, dx_i \tag{1.10}$$

In the case where the only work involves the system expanding against an external pressure, this expression becomes

$$\delta w = -P_{\text{ext}} \, dV' \tag{1.11}$$

This expression occurs in many applications, so other forms of work are referred to as "non-PV work" and are represented by $\delta w'$. Thus the total work done on the system is written as

$$\delta w = -P_{\text{ext}} \, dV' + \delta w' \tag{1.12}$$

An important example of non-PV work is *surface work*, which will be introduced in Chapter 2 and will play a major role in most of this book.

1.1.2.2 Heat

Heat is energy, which is in motion under the influence of a driving force, which is a temperature difference. Two important special cases are heat transport under conditions of constant volume and constant pressure. In the former case, in the absence of $\delta w'$, the path-dependent function q becomes the change in the state function E, following Equation (1.9).

Under these conditions the parameter relating q to the temperature difference is the constant-volume heat capacity,

$$C_V \equiv \left(\frac{\partial E}{\partial T}\right)_V \tag{1.13}$$

Thus, the heat flow associated with changing the temperature of a system from T_1 to T_2 is

$$q = \Delta E' = n \int_{T_1}^{T_2} C_V \, dT \tag{1.14}$$

The constant-pressure case is conveniently formulated in terms of the state function known as the *enthalpy*:

$$H' \equiv E' + PV' \tag{1.15}$$

The first law in terms of enthalpy becomes

$$dH' = dE' + P\,dV' + V'\,dP = q + V'\,dP + \delta w' \tag{1.16}$$

In the case of constant pressure and the absence of $\delta w'$, the path-dependent function q becomes the change in the state function H. Under these conditions the parameter relating q to the temperature difference is the constant-pressure heat capacity,

$$C_P \equiv \left(\frac{\partial H}{\partial T}\right)_P \tag{1.17}$$

Thus, the heat flow associated with changing the temperature of a system from T_1 to T_2 at constant pressure is

$$q = \Delta H' = n \int_{T_1}^{T_2} C_P \, dT \tag{1.18}$$

1.1.3 The second law

The question of whether or not a process can occur is answered by the *second law of thermodynamics*. The mathematical statement of the second law, which follows directly from the empirical statements, is, for an isolated system,

$$dS' \geq 0 \quad (1.19)$$

where S is the state function known as *entropy*, which is defined by

$$dS' \equiv \frac{\delta q_{\text{rev}}}{T} \quad (1.20)$$

The inequality in Equation (1.19) pertains to a process that will tend to occur irreversibly (spontaneously), whereas the equality pertains to a *reversible process*, i.e. one in which the system is never displaced from equilibrium by a finite amount. An equivalent expression for the second law, which does not require the constraint of an isolated system, is

$$dS'_{\text{system}} + dS'_{\text{surroundings}} \geq 0 \quad (1.21)$$

The surroundings are presumed to behave reversibly, so that Equation (1.21) may be written as

$$dS'_{\text{system}} + \frac{\delta q_{\text{surroundings}}}{T} \geq 0 \quad (1.22)$$

or, noting that $\delta q_{\text{surroundings}} = -\delta q_{\text{system}}$,

$$\delta q_{\text{system}} \leq T \, dS'_{\text{system}} \quad (1.23)$$

Note that, for the special case of constant pressure and $\delta w' = 0$, Equation (1.20) may be written as

$$dS' = \frac{dH'}{T} = \frac{nC_P \, dT}{T} \quad (1.24)$$

1.1.4 The third law

The third law is based on the observation that the entropy change for some reactions approaches zero as the temperature approaches 0 K. If the entropy of the component elements is arbitrarily set to zero then the entropy of a compound formed from those elements would also be zero. This has been discussed in detail by Lupis in Section I.3 of Reference [3]. The third law may then be expressed as follows:

If the entropy of each element in complete thermodynamic equilibrium is taken as zero at zero Kelvin the entropy of every other substance becomes zero at zero Kelvin if the substance is in complete thermodynamic equilibrium.

In equation form this may be expressed as

$$S_{0K}^\circ = 0 \tag{1.25}$$

and the entropy of the substance at any temperature, T, is just the entropy increment for heating the substance from 0 K to T, i.e. absolute values of the entropy may be calculated.

1.1.5 Combined first and second laws

Many of the useful applications of thermodynamics result from combining the first and second laws in terms of appropriate functions under the assumption of reversible conditions. The second law may be combined with the first law by substituting Equation (1.23) for the heat absorbed by the system into Equation (1.9), giving

$$dE' \leq T\,dS' - P\,dV' + \delta w' \tag{1.26}$$

The combined first and second laws may also be written in terms of other state functions, which are defined for convenience in solving certain types of problems. These include the *enthalpy* (Equation (1.15)) such that

$$dH' = dE' + P\,dV' + V'\,dP \leq T\,dS' + V'\,dP + \delta w' \tag{1.27}$$

the *Helmholtz free energy*,

$$F' \equiv E' - TS' \tag{1.28}$$

which yields

$$dF' \leq -P\,dV' - S'\,dT + \delta w' \tag{1.29}$$

and the *Gibbs free energy*,

$$G' \equiv H' - TS' \tag{1.30}$$

which yields

$$dG' \leq V'\,dP - S'\,dT + \delta w' \tag{1.31}$$

Two useful relations result from Equations (1.29) and (1.31). Firstly, under isothermal conditions Equation (1.29) becomes

$$dF' \leq -P\,dV' + \delta w' = \delta w_{\text{tot}} = -\delta w_{\text{by system}} \tag{1.32}$$

The relation in Equation (1.32) indicates that the work done by the system will always be less than or equal to the negative of the change in Helmholtz free energy. Thus the maximum isothermal work which can be obtained from the system will correspond to the equality, i.e. reversible conditions. Similarly under isothermal, isobaric conditions Equation (1.31) becomes

$$dG' \leq \delta w' \tag{1.33}$$

Multicomponent and open systems require additional terms in Equations (1.26), (1.27), (1.29) and (1.31) to include the contributions to the various functions made by adding or removing matter from the system. These terms are the chemical potentials of each component in the system, defined by

$$\mu_i \equiv \left(\frac{\partial E'}{\partial n_i}\right)_{S',V',n_1,n_2,\ldots} \tag{1.34}$$

Thus, Equation (1.26) becomes

$$dE' \leq T\,dS' - P\,dV' + \sum_{i=1}^{N} \mu_i\, dn_i + \delta w' \qquad (1.35)$$

for a multicomponent system. Addition of the identity

$$d(PV' - TS') = P\,dV' + V'\,dP - T\,dS' - S'\,dT \qquad (1.36)$$

to Equation (1.35) yields

$$dG' \leq V'\,dP - S'\,dT + \sum_{i=1}^{N} \mu_i\, dn_i + \delta w' \qquad (1.37)$$

Thus, the chemical potential of a component is equivalent to its *partial molar Gibbs free energy*,

$$\mu_i = \left(\frac{\partial G'}{\partial n_i}\right)_{T,P,n_1,n_2,\ldots} = \bar{G}_i \qquad (1.38)$$

Similar operations yield

$$\mu_i = \left(\frac{\partial H'}{\partial n_i}\right)_{S',P,n_1,n_2,\ldots} = \left(\frac{\partial F'}{\partial n_i}\right)_{T,V',n_1,n_2,\ldots} \qquad (1.39)$$

It should be noted that the chemical potentials written in terms of E', H' and F' do not correspond to partial molar quantities.

1.1.5.1 A note on Maxwell reciprocal relations

The combined first and second laws all have the same mathematical form. If reversible conditions are invoked, the general form is that of a perfect differential,

$$dQ' = \left(\frac{\partial Q'}{\partial x_1}\right)_{x_2,x_3,\ldots} dx_1 + \left(\frac{\partial Q'}{\partial x_2}\right)_{x_1,x_3,\ldots} dx_2 + \left(\frac{\partial Q'}{\partial x_3}\right)_{x_1,x_2,\ldots} dx_3 + \cdots$$

$$(1.40)$$

where x_1, x_2, x_3, etc. are appropriate state variables. The mathematical properties of Equation (1.40) are such that the mixed second partial derivatives of Q' are independent of the order of differentiation. For example,

$$\left(\frac{\partial^2 Q'}{\partial x_1 \, \partial x_2}\right) = \left(\frac{\partial^2 Q'}{\partial x_2 \, \partial x_1}\right) \tag{1.41}$$

where all variables other than the differentiation variable are held constant in each operation. Equation (1.41) is the general form of the *Maxwell reciprocal relations*, which are useful for implementing changes of variable in various calculations. As an example, if $\delta w' = 0$ in Equation (1.37) and T and P are the variables of differentiation, Equation (1.41) becomes

$$\left(\frac{\partial V}{\partial T}\right)_{P,n_i} = -\left(\frac{\partial S}{\partial P}\right)_{T,n_i} \tag{1.42}$$

This expression is useful in that it converts the pressure dependence of entropy to a simple quantity,

$$\left(\frac{\partial V}{\partial T}\right)_{P,n_i}$$

which is directly related to the coefficient of thermal expansion. Clearly there are many Maxwell relations, which arise from the different forms of the combined first and second laws depending on the operative system state variables. A more extensive discussion of these relations is presented in Chapter 5 of Reference [2].

1.2 Multicomponent systems – solution thermodynamics

The changes in properties when a solution is formed from its components are described by the mixing quantities. The partial molar mixing quantities for each component are given by

$$\Delta \bar{Q}_i^M = \bar{Q}_i - Q_i^\circ \tag{1.43}$$

where Q_i° is the molar quantity for unmixed component i. The molar mixing quantity is given by

$$\Delta Q^M = Q - \sum_i X_i Q_i^\circ = \sum_i (\bar{Q}_i - Q_i^\circ) \quad (1.44)$$

In multicomponent systems the partial Gibbs free energy of mixing and chemical potential of a given component are related to composition via its *activity* by expressions of the type

$$\Delta \bar{G}_A^M = \mu_A - \mu_A^\circ = RT \ln a_A \quad (1.45)$$

and

$$\Delta \bar{G}_B^M = \mu_B - \mu_B^\circ = RT \ln a_B \quad (1.46)$$

where μ_i° is the chemical potential of component i ($i = A, B, \ldots$) in a defined standard state (often taken as pure component i under conditions of $P = 1$ atm and the temperature of interest) and a_i is the thermodynamic activity which describes the deviation from the standard state for a given species and may be expressed for a given component i as

$$a_i = \frac{f_i}{f_i^\circ} \approx \frac{p_i}{p_i^\circ} \quad (1.47)$$

where f_i is either the fugacity in the vapor phase over a condensed species or the fugacity of a gaseous species and p_i represents the corresponding partial pressures. The terms f_i° and p_i° are the same quantities but corresponding to the standard state of i. Expressing a_i by the partial pressures in Equation (1.47) requires the reasonable approximation of ideal-gas behavior at the high temperatures and relatively low pressures usually encountered. The activity varies with composition.

1.2.1 The ideal-solution model

The simplest model for describing solution behavior is that for an *ideal solution* that obeys *Raoult's law*,

$$a_i = X_i \tag{1.48}$$

The insertion of Raoult's law into the relevant thermodynamic functions results in the following mixing expressions for the simple case of a binary solution A–B:

$$\Delta G^{M,id} = RT \left[X_A \ln X_A + X_B \ln X_B \right] \tag{1.49}$$

$$\Delta S^{M,id} = -R \left[X_A \ln X_A + X_B \ln X_B \right] \tag{1.50}$$

$$\Delta H^{M,id} = 0 \tag{1.51}$$

$$\Delta V^{M,id} = 0 \tag{1.52}$$

The ideal-solution model is greatly oversimplified, but it provides a good basis for comparing the behaviors of real solutions.

1.2.2 Non-ideal solutions

Figure 1.2 shows activity versus composition plots for simple non-ideal binary solutions, with Figure 1.2(a) corresponding to a slightly positive deviation from Raoult's law ($a > X$) and Figure 1.2(b) corresponding to a slightly negative deviation ($a < X$). The relation of activity to concentration is quantified through the activity coefficient, γ_i. When mole fraction is chosen as the concentration unit this relation is given by

$$a_i = \gamma_i X_i \tag{1.53}$$

Figure 1.2 indicates that the activity of each component approaches Raoult's law as its mole fraction approaches unity. It also indicates

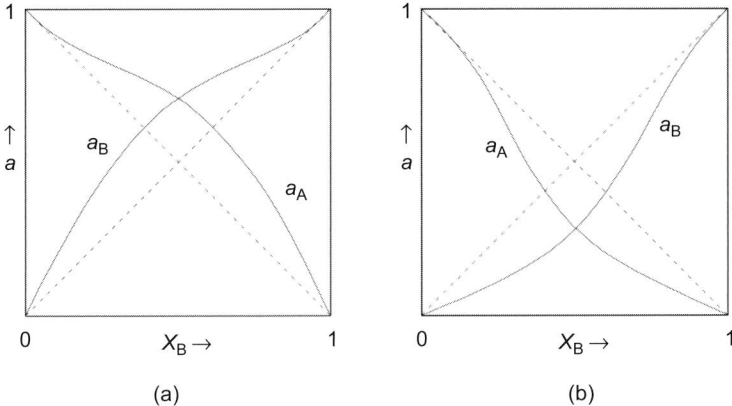

Figure 1.2 Activity versus composition diagrams for a solution exhibiting (a) positive deviation from Raoult's law and (b) negative deviation from Raoult's law.

that there is a linear dependence of activity on concentration at low concentration. In this region, γ_i assumes a constant value, γ_i°, and the relation between activity and concentration becomes

$$a_i = \gamma_i^\circ X_i \tag{1.54}$$

This expression is known as *Henry's law*.

The simplest model for a non-ideal solution is the *regular solution*. This model will be reviewed here because the concepts underlying the model will provide a convenient method for estimating surface quantities in subsequent chapters. Consider a solution formed from components A and B for which the coordination number of the atoms, Z, is the same in both components and in the solution. The internal energies of the pure components will be given by the number of bonds between atoms multiplied by the respective bond energies, ε_{ii},

$$E_A^\circ = N_{AA}^\circ \varepsilon_{AA} = \tfrac{1}{2} N_A Z \varepsilon_{AA} \tag{1.55}$$

$$E_B^\circ = N_{BB}^\circ \varepsilon_{BB} = \tfrac{1}{2} N_B Z \varepsilon_{BB} \tag{1.56}$$

where N_{AA}° and N_{BB}° are the numbers of bonds and N_A and N_B are the numbers of atoms. The physical significance of the bond energies may be seen by considering a process whereby the atoms from the condensed phase are widely separated into the vapor phase so that they no longer interact, i.e. $E = 0$. If the component is a liquid and the approximation $H \approx E$ is made, the enthalpy change for this process will be the molar *heat of vaporization* if Avogadro's number, N_0, of atoms are separated,

$$\Delta H_A^{vap} = 0 - \tfrac{1}{2} N_0 Z \varepsilon_{AA} \qquad (1.57)$$

Rearrangement of Equation (1.57) yields the bond energy as

$$\varepsilon_{AA} = -\frac{2 \Delta H_A^{vap}}{N_0 Z} \qquad (1.58)$$

Thus the bond energy is a negative quantity and is essentially the negative of the heat of vaporization per bond broken. If A were a solid, the enthalpy change in Equation (1.58) would be the *heat of sublimation*. The bond energies associated with B—B and A—B bonds may be interpreted in a similar way. Therefore, the internal energy of a solution may be calculated by multiplying the number of bonds of each type by the appropriate bond energy. In the simplest case it is assumed that the bond energies are independent of concentration such that ε_{AA} and ε_{BB} are the same in the solution as they are in the unmixed components,

$$E_{Soln} = N_A \varepsilon_{AA} + N_B \varepsilon_{BB} + N_{AB} \varepsilon_{AB} \qquad (1.59)$$

The number of atoms of A is related to the number of bonds by the following expression:

$$N_A = \frac{N_{AB}}{Z} + \frac{2 N_{AA}}{Z} \qquad (1.60)$$

Rearrangement of Equation (1.60) yields the number of A—A bonds as

$$N_{AA} = \tfrac{1}{2}(Z N_A - N_{AB}) \qquad (1.61)$$

A similar expression obtains for the number of B—B bonds,

$$N_{BB} = \tfrac{1}{2}(ZN_B - N_{AB}) \tag{1.62}$$

Substitution of Equations (1.61) and (1.62) into Equation (1.59) allows the internal energy to be expressed according to Equation (1.60):

$$E_{Soln} = \tfrac{1}{2}ZN_A\varepsilon_{AA} + \tfrac{1}{2}ZN_B\varepsilon_{BB} + N_{AB}\left(\varepsilon_{AB} - \frac{\varepsilon_{AA} + \varepsilon_{BB}}{2}\right) \tag{1.63}$$

Comparison with Equations (1.55) and (1.56) indicates that the first two terms on the right-hand side of Equation (1.63) are the internal energies of the unmixed components. Therefore, the third term is the internal energy of mixing,

$$\Delta E^M = N_{AB}\left(\varepsilon_{AB} - \frac{\varepsilon_{AA} + \varepsilon_{BB}}{2}\right) \tag{1.64}$$

It remains to express the number of A—B bonds, N_{AB}, as a function of the composition of the solution. Assuming one mole of solution, N_0 atoms, the total number of bonds will be

$$N_{tot} = \frac{N_0 Z}{2} \tag{1.65}$$

The number of A—B bonds will be this number multiplied by the fraction of bonds that are of the type A—B, i.e. the probability that a given bond is of the type A—B. The regular solution assumes random mixing of the atoms, which facilitates the calculation of this fraction. Consider a single bond between two sites in the solution, Figure 1.3(a). The mathematical probability that site 1 will be occupied by an A atom is just X_A and the probability that site 2 will be occupied by a B atom is X_B. Thus the probability that the bond is an A—B bond, Figure 1.3(b), is $X_A X_B$. However, an A—B bond could also be obtained by the site occupancy in Figure 1.3(c). for which the probability is $X_B X_A$. Therefore

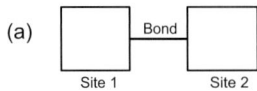
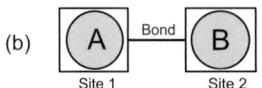
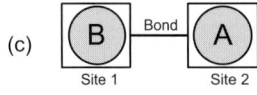

Figure 1.3 A schematic diagram of the bonding between two atoms in solution. (a) Two empty adjacent sites. (b) The same two sites, with site 1 occupied by an A atom and site 2 occupied by a B atom. (c) The same sites, with site 1 occupied by a B atom and site 2 occupied by an A atom.

the probability that the given bond is an A—B bond is $2X_A X_B$ and the number of A—B bonds will be given by

$$N_{AB} = \frac{N_0 Z}{2}(2X_A X_B) \qquad (1.66)$$

The substitution of N_{AB} into Equation (1.64) gives the following expression for ΔE^M, which for solid or liquid solutions is virtually equal to ΔH^M:

$$\Delta H^{M,\text{reg}} \approx \Delta E^{M,\text{reg}} = N_0 Z \left(\varepsilon_{AB} - \frac{\varepsilon_{AA} + \varepsilon_{BB}}{2} \right) X_A X_B = \Omega X_A X_B \qquad (1.67)$$

The relative strengths of the bond energies for like and unlike atoms will determine the sign and the magnitude of the deviation from ideal-solution behavior. Since the regular-solution model assumes random mixing of the atoms, the entropy of mixing will be the same as that for an ideal solution so the Gibbs free energy of mixing will be given by

$$\Delta G^{M,\text{reg}} = \Omega X_A X_B + RT[X_A \ln X_A + X_B \ln X_B] \qquad (1.68)$$

Addition of the free energies of the unmixed components to Equation (1.68) yields the molar free energy of the solution:

$$G^{\text{reg}} = X_A \mu_A^\circ + X_B \mu_B^\circ + \Omega X_A X_B + RT[X_A \ln X_A + X_B \ln X_B] \qquad (1.69)$$

The approach used in developing the regular-solution model, particularly the assumption of random mixing of interacting atoms, is a great oversimplification. Clearly systems for which Ω is negative will try to maximize the number of bonds between unlike atoms, i.e. they will tend to order. Similarly, systems for which Ω is positive will try to maximize the number of bonds between like atoms, i.e. they will tend to cluster. More sophisticated bond-breaking models are available – see Chapters XV and XVI of Reference [3]. In this book the regular model will generally be used to illustrate trends in behavior in a simple manner. (Note that in this form the regular-solution model is also called the *quasichemical solution model*.)

1.3 Multiphase equilibria

The behavior of multiphase systems is described by the use of *phase diagrams*. The following is a review of important terminology.

Phase diagram – a "map" that indicates which phases are stable as a function of the relevant thermodynamic variables (e.g. T, P, composition, etc.) for a given system.

Phase – a physically distinct part of a system whose intensive properties are homogeneous and which has definite bounding surfaces.

Homogeneous system – a system that consists of one phase, e.g. water in a glass.

Heterogeneous system – a system that consists of two or more phases, e.g. water (liquid phase) and ice cubes (solid phase) in a glass.

Number of components – the smallest number of substances whose concentrations define the composition of each part of the system (phase).

Unary = one component
Binary = two components
Ternary = three components

Number of degrees of freedom – the number of externally controllable intensive variables (T, P, concentration, etc.) which can be independently altered without bringing about the disappearance of a phase or the appearance of a new phase.

Invariant = zero degrees of freedom
Univariant = one degree of freedom
Bivariant = two degrees of freedom

Depending on the nature of the system and the prevailing experimental variables, there is an almost infinite number of possible phase diagrams. However, each of the diagrams falls into one of three classes based on the type of variables used to plot the diagram [6]. The type of variables may be defined relative to the combined first and second laws written in terms of the internal energy, Equation (1.35). With $\delta w'$ taken as zero the coefficients of the remaining terms (T, $-P$ and μ_i) are *potentials*, "defined as the rate of change of internal energy with the conjugate extensive variable (S', V' and n_i)" [6]. The three general types of phase diagram are

Type 1 – a plot of two potentials;
Type 2 – a plot of one potential and one ratio of conjugate extensive quantities; and
Type 3 – a plot of two ratios of conjugate extensive quantities.

Each type of diagram has common features as described below.

1.3.1 Unary systems

Figure 1.4 presents the phase equilibria in a hypothetical unary system represented on each of the three types of diagrams. The plot of P versus

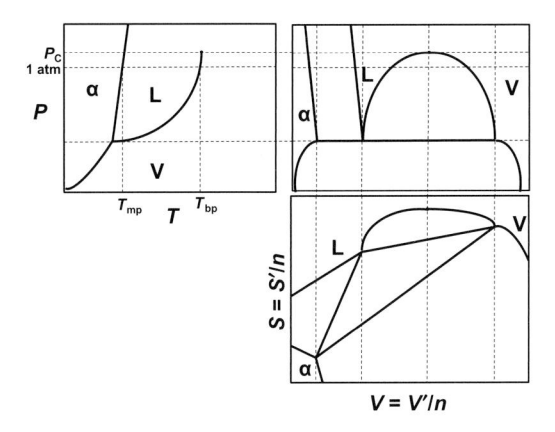

Figure 1.4 Phase diagrams for a hypothetical unary system in the form of a Type 1 (*P* versus *T*) diagram, a Type 2 (*P* versus *V*) diagram and a Type 3 (*S* versus *V*) diagram.

T is a Type 1 diagram and the three-phase equilibrium α–L–V is represented by a point, the *triple point*. The plot of *P* versus *V* is a Type 2 diagram and the α–L–V equilibrium is represented by three points on a line. The plot of *S* versus *V* (two molar quantities) is a Type 3 diagram and the α–L–V equilibrium is represented by an area. (Note that the three-phase equilibria are represented in these ways on all phase diagrams, on the basis of their "Type" irrespective of whether they are unary, binary, etc.)

The horizontal line on the *P*–*T* diagram at 1 atm crosses the α–L equilibrium line at the normal melting point and crosses the L–V equilibrium line at the normal boiling point. Application of the second law to these equilibria can be done conveniently with the *G* versus *T* plot in Figure 1.5. The melting point is the temperature at which

$$\Delta G_{T,P}(\alpha \to L) = 0 \tag{1.70}$$

Therefore, this temperature is that at which the *G* versus *T* curves for the liquid and solid phases cross. Similarly, the normal boiling point is the temperature at which the curves for the vapor and liquid phases cross.

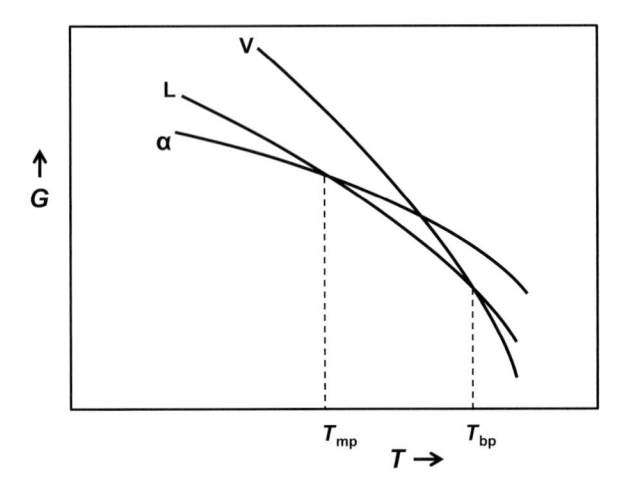

Figure 1.5 A schematic plot of G versus T at a pressure of 1 atm for the unary system in Figure 1.4.

1.3.1.1 The Gibbs phase rule for unary systems

The *Gibbs phase rule* is a fundamental underlying principle, which must be followed by any correct phase diagram. The following is a derivation of this principle for a unary system. Let φ represent the number of phases (I, II, III) and let f represent the number of degrees of freedom. Two of the three variables T, P, V are required in order to specify the state of each phase. (Generally T and P are taken as the independent variables.) Thus, the number of variables needed to specify the state of the entire system is 2φ. However, many of the variables are fixed by the conditions of equilibrium: thermal equilibrium gives $T^I = T^{II} = T^{III} = \ldots$, i.e. ($\varphi - 1$) equations; mechanical equilibrium gives $P^I = P^{II} = P^{III} = \ldots$, i.e. ($\varphi - 1$) equations; and thermodynamic equilibrium gives $G^I = G^{II} = G^{III} = \ldots$, i.e. ($\varphi - 1$) equations.

This results in a total of $3(\varphi - 1)$ equations. The number of degrees of freedom (variance) will be the difference between the number of variables and the number of equations relating them:

$$f = \text{No. of variables} - \text{No. of equations} = 2\varphi - 3(\varphi - 1) \quad (1.71)$$

or

$$f = 3 - \varphi \tag{1.72}$$

Equation (1.72) is the unary *Gibbs phase rule*. It indicates that the maximum number of phases which can coexist in a unary system is 3 and this results in an invariant equilibrium ($f = 0$). Note that the equilibria in each type of phase diagram in Figure 1.4 satisfy this condition.

1.3.2 Multicomponent systems

The equilibrium state of a system at constant temperature and pressure is characterized by a minimum in the Gibbs free energy of the system according to the second law of thermodynamics. For a multicomponent, multiphase (bulk) system, the minimum free energy corresponds to uniformity of the chemical potential (μ) of each component throughout the system as demonstrated below. For a binary system, the molar free energy (G) and chemical potentials are related by

$$G = (1 - X)\mu_A + X\mu_B \tag{1.73}$$

where μ_A and μ_B are the chemical potentials of components A and B, respectively, and $(1 - X)$ and X are the mole fractions of A and B. Figure 1.6(a) is a plot of G versus composition for a single phase, which exhibits simple solution behavior. For this case Equations (1.7) and (1.8) become

$$\mu_A = G - X\frac{dG}{dX} \tag{1.74}$$

$$\mu_B = G + (1 - X)\frac{dG}{dX} \tag{1.75}$$

which, as indicated in Figure 1.6(a), mean that a tangent drawn to the free-energy curve has intercepts on the ordinate at $X = 0$ and $X = 1$ that correspond to the chemical potentials of A and B, respectively.

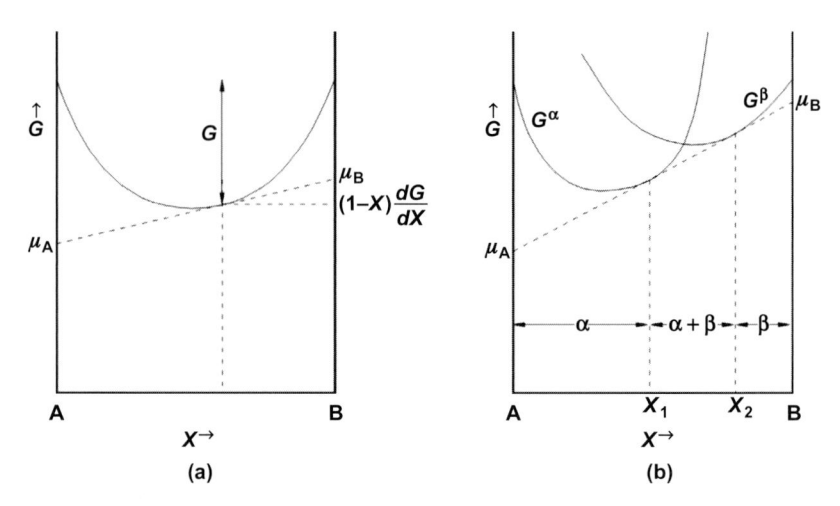

Figure 1.6 Free-energy versus composition diagrams for (a) one phase and (b) two coexisting phases in a binary system.

Figure 1.6(b) shows the free energy versus composition diagram for two phases. Application of the lever rule shows that the free energy for a mixture of the two phases is given by the point where the chord drawn between the two individual free-energy curves intersects the bulk composition. Clearly, for any bulk composition between X_1 and X_2 in Figure 1.6(b), a two-phase mixture will have a lower free energy than a single solution corresponding to either phase, and the lowest free energy of the system, i.e. the lowest chord, obtains when the chord is in fact a tangent to both free-energy curves. This "*common tangent construction*" indicates that X_1 and X_2 are the equilibrium compositions of the two coexisting phases and the intercepts of the tangent at $X = 0$ and $X = 1$ demonstrate the equality of chemical potentials, i.e.

$$\mu_A^\alpha = \mu_A^\beta \tag{1.76}$$

$$\mu_B^\alpha = \mu_B^\beta \tag{1.77}$$

The reader is referred to works by Hillert [5, 7] for a comprehensive treatment of free-energy diagrams and their applications to phase transformations.

Alternatively, if the free energy of the unmixed components in their standard states, $(1 - X)\mu_A^\circ + X\mu_B^\circ$, is subtracted from each side of Equation (1.73), the result is the molar free energy of mixing

$$\Delta G^M = (1 - X)(\mu_A - \mu_A^\circ) + X(\mu_B - \mu_B^\circ) \qquad (1.78)$$

and a plot of ΔG^M versus X may be constructed, in which case the tangent intercepts are $(\mu_A - \mu_A^\circ)$ and $(\mu_B - \mu_B^\circ)$, respectively.

Equations (1.76) and (1.77), together with Equations (1.45) and (1.46), indicate that similar information may be expressed in terms of activity. For cases where a two-phase mixture is stable, (Figure 1.6(b)), Equations (1.76) and (1.77) indicate that the activities are constant across the two-phase field. Figure 1.7 illustrates the above points and their relationship to the conventional T versus X phase diagram for a hypothetical binary eutectic system.

Clearly, if the variation of G or ΔG^M with composition for a system can be established, either by experimental activity measurements or by calculations using solution models, the phase equilibria can be predicted for a given temperature and composition. Considerable effort has been spent in recent years on computer calculations of phase diagrams. Furthermore, the concepts need not be limited to binary elemental systems. For example, ternary free-energy behavior may be described on a three-dimensional plot, the base of which is a Gibbs triangle and the vertical axis represents G (or ΔG^M). In this case, the equilibrium state of the system is described using a *"common tangent plane"*, rather than a common tangent line, but the principles are unchanged. Systems with more than three components may be treated, but graphical interpretation becomes more difficult.

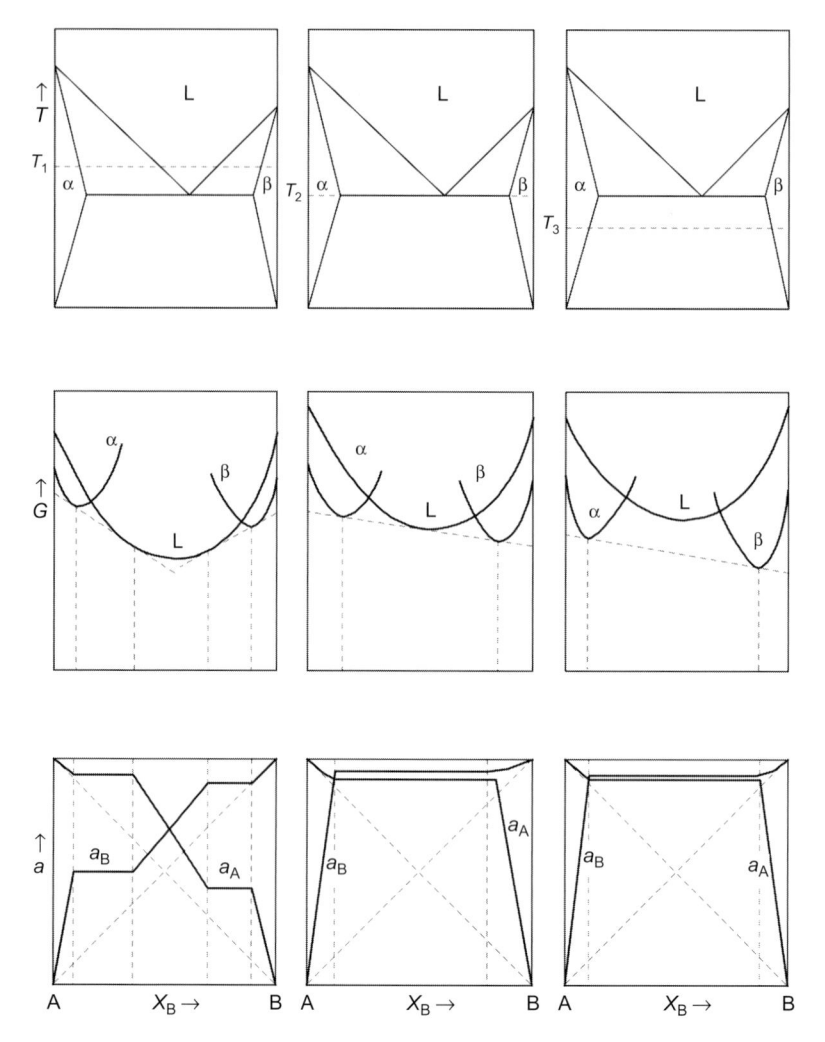

Figure 1.7 Free-energy versus composition and activity diagrams for a hypothetical binary eutectic system.

1.3.2.1 The Gibbs phase rule

The application of the *Gibbs phase rule* to multicomponent systems containing c components may be done by extending the treatment used for the unary system in Section 1.2.1.1. Again let φ represent the number of phases (I, II, III) and let f represent the number of degrees of freedom. Now $(c-1)$ compositions are required, in addition to two of the three variables T, P, V, which are required in order to specify the state of each phase. Thus, the number of variables needed to specify the state of each phase is $(c+1)$ and the number to specify the state of the entire system is $(c+1)\varphi$. The variables which are fixed by the conditions of equilibrium are as follows: thermal equilibrium gives $T^{I} = T^{II} = T^{III} = \ldots$, i.e. $(\varphi-1)$ equations; mechanical equilibrium gives $P^{I} = P^{II} = P^{III} = \ldots$, i.e. $(\varphi-1)$ equations; and thermodynamic equilibrium gives $\mu_{A}^{I} = \mu_{A}^{II} = \mu_{A}^{III} = \ldots$, i.e. $(\varphi-1)$ equations, $\mu_{B}^{I} = \mu_{B}^{II} = \mu_{B}^{III} = \ldots$, i.e. $(\varphi-1)$ equations and $\mu_{C}^{I} = \mu_{C}^{II} = \mu_{C}^{III} = \ldots$, i.e. $(\varphi-1)$ equations, etc.

The thermodynamic equilibrium conditions interrelate concentrations and temperature through equations such as Equations (1.45) and (1.53) and contribute $c(\varphi-1)$ equations. The equilibrium conditions together result in a total of $(c+2)(\varphi-1)$ equations. The number of degrees of freedom is again the difference between the number of variables and the number of equations relating them:

$$f = \text{No. of variables} - \text{No. of equations} = (c+1)\varphi - (c+2)(\varphi-1)$$

(1.79)

or

$$f = c\varphi + \varphi - c\varphi + c - 2\varphi + 2 \qquad (1.80)$$

Equation (1.80) simplifies to

$$f = c - \varphi + 2 \qquad (1.81)$$

Equation (1.81) is the generalized *Gibbs phase rule*. Note that Equation (1.81) reduces to Equation (1.72) for the special case of a unary system.

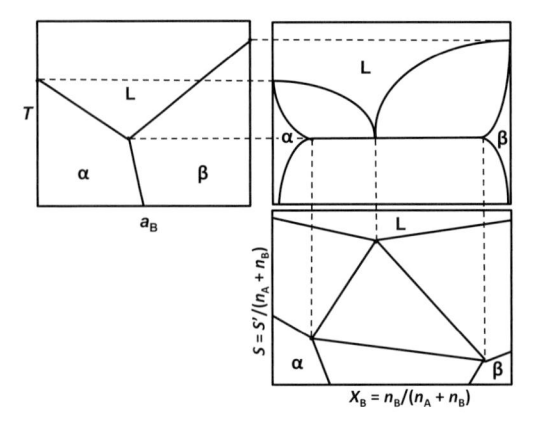

Figure 1.8 Phase diagrams for a hypothetical binary system in the form of a Type 1 (T versus a_B) diagram, a Type 2 (T versus X_B) diagram and a Type 3 (S versus X_B) diagram.

1.3.2.2 Simple phase diagrams

Figure 1.8 presents the phase equilibria in a hypothetical binary eutectic system similar to that in Figure 1.7, represented on each of the three types of diagrams. This diagram is similar to those for the Ag–Cu and Ni–Cr systems. The plot of T versus a_B is a Type 1 diagram and the three-phase equilibrium α–L–β is represented by a point. The plot of T versus X_B is a Type 2 diagram and the α–L–β equilibrium is represented by three points on a line, the eutectic isotherm. The plot of S versus X_B is a Type 3 diagram and the α–L–β equilibrium is represented by an area. Note that the forms of these diagrams correspond to those for the unary system in Figure 1.4. (Numerous examples of the three types of phase diagrams are given for unary, binary and ternary systems in Chapter 13 of Reference [2], Reference [5] and Chapter 2 of Reference [8].

Figure 1.9 is a temperature versus composition diagram for a hypothetical binary system with a minimum in the solidus and liquidus curves and a miscibility gap in the solid state. This is similar to the binary

1.3 MULTIPHASE EQUILIBRIA 27

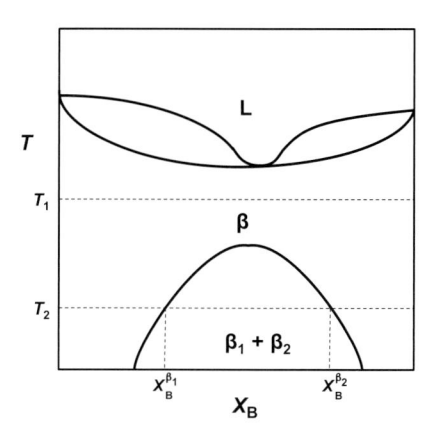

Figure 1.9 A temperature versus composition diagram for a hypothetical binary system with a minimum in the solidus and liquidus curves and a miscibility gap in the solid state.

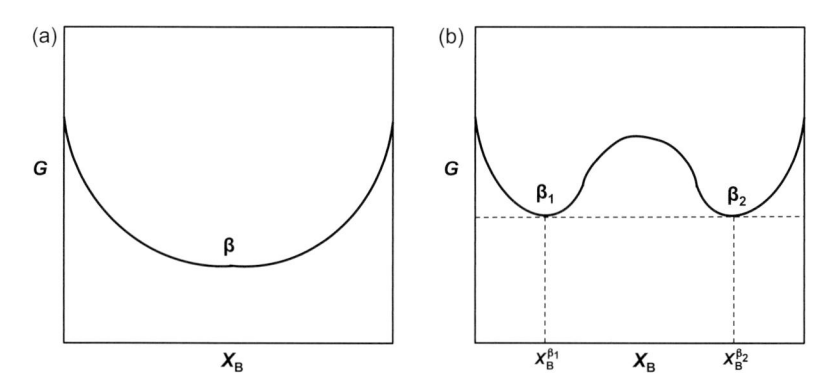

Figure 1.10 G versus X_B plots at two temperatures for the phase diagram in Figure 1.9: (a) T_1, which is above the critical point of the miscibility gap; and (b) T_2, which is below the critical point. (The free-energy curve for the liquid phase is not shown.)

Au–Ni system. Clearly the phases β_1 and β_2 must have the same crystal structure to allow the continuous series of solid solutions above the critical point of the miscibility gap. Figure 1.10 presents G versus X_B plots at two temperatures: T_1, which is above the critical point of the miscibility gap; and T_2, which is below. The free-energy curve for the liquid phase is not shown. The free-energy curve at T_1 is everywhere concave upward, which is consistent with a single phase, β, being stable. At T_2 the curve has developed two inflections, resulting in two "lobes" that allow a common tangent to be drawn between them. The two points of tangency indicate the equilibrium compositions of phases $β_1$ and $β_2$ at T_2. A phase diagram such as that in Figure 1.9 results from a situation where the liquid phase exhibits near-ideal-solution behavior and the solid phase exhibits a significant positive deviation from ideality. Such a diagram can be calculated by assuming two regular solutions with $Ω_L ≈ 0$ and $Ω_S$ positive. See Problem 9.

1.3.2.3 Metastability

The discussion of phase equilibria to this point has focused upon stable equilibria, i.e. absolute minima in the Gibbs free energy. However, many important technological systems rely on phases that are in metastable equilibria, i.e. they correspond to relative minima in G. These include cementite (Fe_3C) in carbon steels, GP zones in age-hardening Al-base alloys, the t′ phase in ZrO_2–Y_2O_3 ceramics, etc. The laws of thermodynamics and the equilibrium conditions may be applied to such systems if the "lifetimes" of the phases are long relative to the time of observation. Figure 1.11 presents a simple phase diagram on which the equilibrium solid phases are α and $β_1$, with the phase boundaries corresponding to stable equilibrium drawn as solid lines. Also included are the dashed phase boundaries associated with metastable equilibria involving a B-rich solid phase $β_2$, which has a different crystal structure from $β_1$. The diagram indicates the following.

1. The melting point of $β_2$ is lower than that of $β_1$.
2. The solubility of $β_2$ in the liquid is increased over that of $β_1$.
3. The solubility of $β_2$ in α is increased over that of $β_1$.
4. The metastable eutectic equilibium occurs at a lower temperature than the stable eutectic isotherm.

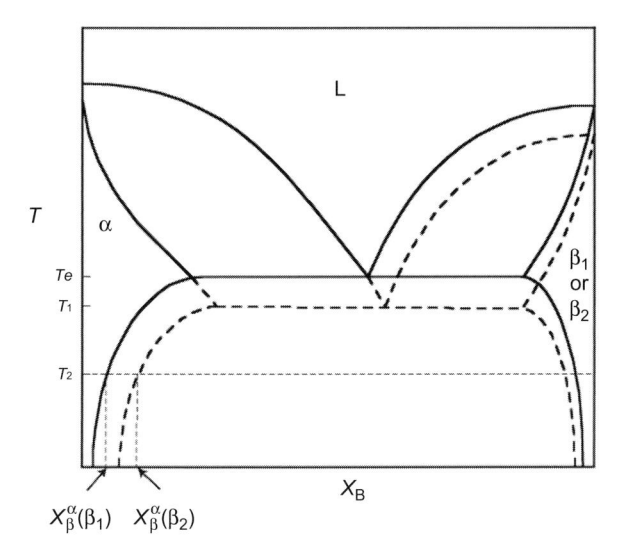

Figure 1.11 A phase diagram with equilibrium solid phases α and β_1 and a metastable B-rich solid phase β_2, which has a different crystal structure from β_1.

Figure 1.12 presents G versus X_B curves at the temperatures indicated in Figure 1.11. The position of the curve for β_2 relative to that for β_1 indicates the cause of the above observations. For example, the increased solubility of β_2 in α is the result of the common tangent between β_2 and α being "steeper" than that between β_1 and α such that it touches the free-energy curve for α at a higher concentration of component B. This is the universal observation that a metastable phase will always have a higher solubility than a stable phase. A discussion of a number of technologically important metallic systems in which metastability plays a significant role may be found in Chapter 5 of Reference [9].

The free energies of phases, such as those in Figures 1.4, 1.7, 1.8 and 1.9, can be shifted by contributions from surfaces and interfaces for phases, which are small in extent. These alterations in the phase equilibria are analogous to those described here for metastability. These important effects will be described in detail in Chapter 6.

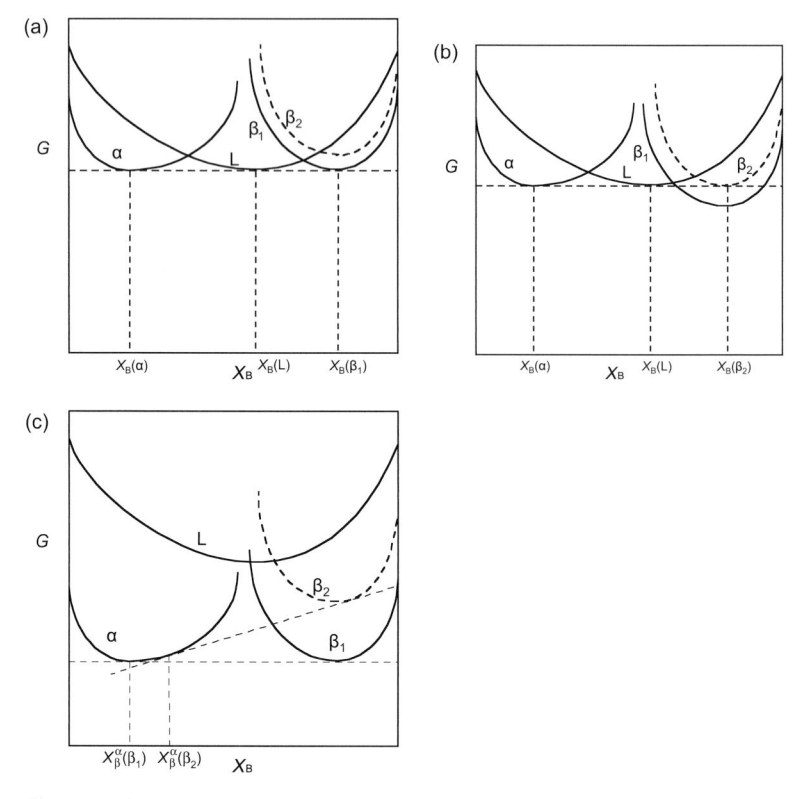

Figure 1.12 G versus X_B curves at the temperatures indicated on the phase diagram in Figure 1.11: (a) T_e, (b) T_1 and (c) T_2.

1.4 Chemical reactions

Since the conditions most often encountered in chemical reactions are constant temperature and pressure, the second law is most conveniently written in terms of the Gibbs free energy (G') of a system. Under these conditions the second law states, following Equation (1.31), that the free-energy change of a process will have the following significance:

$\Delta G' < 0$ spontaneous reaction expected
$\Delta G' = 0$ equilibrium
$\Delta G' > 0$ thermodynamically impossible process

For a generic chemical reaction, e.g.

$$a\text{A} + b\text{B} = c\text{C} + d\text{D} \tag{1.82}$$

where upper-case symbols represent species and lower-case symbols represent the corresponding stoichiometric coefficients. $\Delta G'$ is the difference in free energy between the products and reactants,

$$\Delta G' = c\mu_\text{C} + d\mu_\text{D} - a\mu_\text{A} - b\mu_\text{B} \tag{1.83}$$

and insertion of equations such as Equations (1.45) and (1.46) yields

$$\Delta G' = \Delta G° + RT \ln \left(\frac{a_\text{C}^c a_\text{D}^d}{a_\text{A}^a a_\text{B}^b} \right) \tag{1.84}$$

where $\Delta G°$ is the free-energy change when all species are present in their standard states and a is the activity. The standard free-energy change is expressed for a reaction such as Equation (1.82) by

$$\Delta G° = c\, \Delta G_\text{C}° + d\, \Delta G_\text{D}° - a\, \Delta G_\text{A}° - b\, \Delta G_\text{B}° \tag{1.85}$$

where $\Delta G_\text{C}°$ etc. are standard molar free energies of formation, which may be obtained from tabulated values. (A selected list of references for thermodynamic data such as $\Delta G°$ is included at the end of this chapter – many of these data have also been incorporated into electronic databases used with commercial thermodynamics software.) For the special cases of equilibrium ($\Delta G' = 0$) Equation (1.84) reduces to

$$\Delta G° = -RT \ln \left(\frac{a_\text{C}^c a_\text{D}^d}{a_\text{A}^a a_\text{B}^b} \right)_\text{eq} \tag{1.86}$$

The bracketed term is called the equilibrium constant (K) and is used to describe the equilibrium state of the reaction system,

$$\Delta G° = -RT \ln K \tag{1.87}$$

Equation (1.87) is called the *van 't Hoff isotherm*.

1.4.1 Chemical reactions involving gases

Consider the special case for which the chemical reaction involves oxidation of a metal:

$$M(s) + O_2(g) = MO_2(s) \qquad (1.88)$$

For this case Equation (1.86) becomes

$$\Delta G^{\circ}_{MO_2} = -RT \ln \left(\frac{a_{MO_2}}{a_M p_{O_2}} \right)_{eq} \qquad (1.89)$$

where $\Delta G^{\circ}_{MO_2}$ is the standard Gibbs free energy of formation of MO_2. If the activities of M and MO_2 are taken as unity, Equation (1.89) may be used to express the oxygen partial pressure at which the metal and oxide coexist, i.e. the dissociation pressure of the oxide,

$$p_{O_2}^{M/MO_2} = \exp \left(\frac{\Delta G^{\circ}_{MO_2}}{RT} \right) \qquad (1.90)$$

Often the above information may be conveniently represented in graphical form using Ellingham diagrams, which are plots of the standard free energy of formation (ΔG°) versus temperature for the compounds of a particular type, e.g. oxides, sulfides, carbides, etc. Ellingham diagrams are useful in that they allow comparison of the relative stabilities of various compounds. Figure 1.13 is such a plot for many simple oxides. The values of ΔG° are expressed as kilojoules per mole of O_2, so the stabilities of various oxides may be compared directly, i.e. the lower the position of the line on the diagram the more stable is the oxide.

The values of $p_{O_2}^{M/MO_2}$ may be obtained directly from the oxygen nomograph on the diagram by drawing a straight line from the origin marked O through the free-energy line at the temperature of interest and reading the oxygen partial pressure from its intersection with the

1.4 CHEMICAL REACTIONS 33

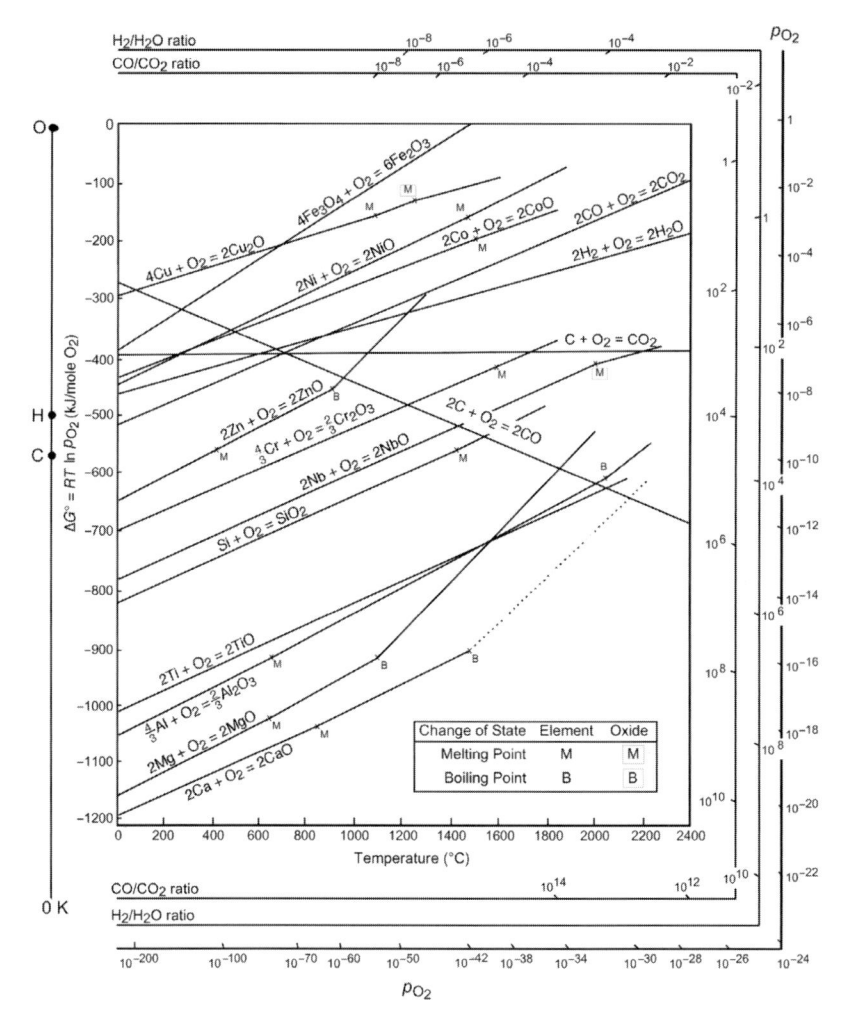

Figure 1.13 An Ellingham–Richardson diagram for oxides.

scale at the right side labeled p_{O_2}. Values for the pressure ratio H_2/H_2O for equilibrium between a given metal and oxide may be obtained by drawing a similar line from the point marked H to the scale labeled "H_2/H_2O ratio" and values for the equilibrium CO/CO_2 ratio may be obtained by drawing a line from point C to the nomograph scale "CO/CO$_2$ ratio". The reader is referred to Chapter 12 of Gaskell [2] for a more detailed discussion of the construction and use of Ellingham diagrams for oxides.

1.5 Summary

This chapter has provided a summary of the laws of thermodynamics and the important defined functions and relationships for applying these laws. The remainder of this book will focus on the way these laws and relationships are influenced by surfaces and interfaces.

1.6 References

[1] L. S. Darken and R. W. Gurry, *Physical Chemistry of Metals* (New York: McGraw-Hill, 1953).
[2] D. R. Gaskell, *Introduction to the Thermodynamics of Materials*, 5th edn. (New York: Taylor and Francis, 2008).
[3] C. H. P. Lupis, *Chemical Thermodynamics of Materials* (New York: Elsevier, 1983).
[4] R. T. DeHoff, *Thermodynamics in Materials Science*, 2nd edn. (New York: Taylor and Francis, 2006).
[5] M. Hillert, The uses of Gibbs free energy–composition diagrams. In *Lectures on the Theory of Phase Transformations*, 2nd edn., ed. H. I. Aaronson (Warrendale, PA: The Minerals, Metals and Materials Society, 1999), pp. 1–33.
[6] A. D. Pelton and H. Schmalzried, On the geometrical representation of phase equilibria, *Metall. Trans.*, **4** (1973), 1395–1404.
[7] M. Hillert, *Phase Equilibria, Phase Diagrams and Phase Transformations*, 2nd edn. (Cambridge: Cambridge University Press, 2008).
[8] N. Birks, G. H. Meier and F. S. Pettit, *Introduction to High Temperature Oxidation of Metals*, 2nd edn. (Cambridge: Cambridge University Press, 2006).

[9] D. A. Porter and K. E. Easterling, *Phase Transformations in Metals and Alloys* (London: Chapman & Hall, 1992).

1.7 Study problems

1. (a) Calculate the enthalpy, entropy, and Gibbs free-energy changes when one mole of supercooled liquid copper solidifies isothermally at 1,000 °C and 1 atm.
   (b) Another one-mole sample of liquid Cu is placed in an adiabatic container at 1,000 °C at 1 atm and solid Cu nucleates. Calculate the entropy change for the ensuing process.
   (c) Show that the second law is obeyed for both of the above processes.
   For Cu, $T_m = 1{,}356$ K, $\Delta H_m = 12{,}970$ J/mole, $C_p(\text{solid}) = 22.64 + 6.28 \times 10^{-3} T$ J/mole K and $C_p(\text{liquid}) = 32.84$ J/mole K.

2. (a) Find the standard entropy of 1 mole of hydrogen at 900 K, $S°_{900}$.
   (b) Find the entropy of 1 mole of hydrogen at 900 K and 0.01 atm. (Assume H_2 is an ideal gas when calculating the change in S with changing pressure.)
   For $H_2(g)$, $S°_{298} = 130.62$ J/mole K, $C_p = 27.3 + 3.3 \times 10^{-3} T + 0.5 \times 10^5 T^{-2}$ J/mole K.

3. Silver exists in the solid state only in the fcc structure which melts at 1,234 K. The boiling point of liquid Ag is 2,147 K.
   (a) Sketch the P–T diagram for silver.
   (b) Sketch a plot of Gibbs free energy versus temperature for the three phases of silver at 1 atm pressure. Indicate important points on your diagram. Explain what determines the slope and curvature of the lines on your diagram. (See Equations (1.24) and (1.30).)
   (c) From the vapor-pressure–temperature relationship for liquid silver below, calculate the heat of evaporation of liquid silver at

the normal boiling point and the heat capacity difference (ΔC_p) between liquid and gaseous silver:

$$\ln p_{Ag} = -\frac{33,200}{T} - 0.85 \ln T + 20.31$$

with p in atm.

4. Assume liquid Au–Cu alloys behave ideally at 1,050 °C.
   (a) Calculate the heat absorbed and the entropy change in the system when 1 mole of solid Cu is dissolved isothermally at this temperature in a large bath of an alloy ($X_{Cu} = 0.50$) Use the data from Problem 1.
   (b) Calculate the heat evolved per mole of Cu oxidized when pure O_2 is blown through the bath described in (a). Au does not oxidize.

   The standard enthalpy of formation of Cu_2O is $\Delta H^\circ_{1,323K} = -318,425$ J/mole.

5. An alloy containing 10 atom percent Ni and 90 atom percent Au is a solid solution at 700 °C. It is found that the solution reacts with water vapor to form NiO. Assume that approximate measurements indicate that the reaction reaches equilibrium when the gas mixture contains 0.35 volume percent hydrogen with the remainder being water vapor. Calculate (a) a_{Ni}, (b) γ_{Ni}, and (c) $\Delta \bar{G}^M_{Ni}$ for this alloy. Does this system exhibit a positive or negative deviation from Raoult's law?

   Take necessary free-energy data from Figure 1.13.

6. Draw schematic free-energy–composition diagrams and plots of activity versus composition at representative temperatures for binary systems A–B that
   (a) form a simple peritectic phase diagram and
   (b) form a phase diagram consisting of two eutectic regions connected by an intermediate phase.
   Re-plot both phase diagrams as T versus a_B.

7. Consider the binary phase diagram (T versus X_B) for the system A–B. Pure A is bcc (α) and melts at 800 K. Pure B is fcc (β) and melts at 1,000 K. There is an intermediate phase (γ) with a tetragonal structure, which has a congruent melting point at 1,100 K and $X_B = 0.5$. There is a peritectic equilibrium at 900 K with $X_B^L = 0.1$, $X_B^\alpha = 0.2$ and $X_B^\gamma = 0.45$. There is a eutectic equilibrium at 850 K with $X_B^\gamma = 0.55$, $X_B^L = 0.7$ and $X_B^\beta = 0.9$.
   (a) Sketch the phase diagram carefully using T–X_B coordinates.
   (b) Use the Gibbs phase rule to calculate the variance of
       (i) the peritectic equilibrium and
       (ii) the congruent melting point.
   Name the relevant components and phases in (i) and (ii).
   (c) Construct schematic plots of ΔG^M versus X_B and a_B versus X_B at 1,000 K and 600 K.
   (d) Redraw the phase diagram using the coordinates temperature versus activity of component B (T versus a_B).

8. Components C and D have different crystal structures and negligible mutual solubility in the solid state. They form a simple eutectic system. For C, $T_m = 1{,}563$ K and $\Delta H_m = 58{,}160$ J/mole. For D, $T_m = 1{,}691$ K and $\Delta H_m = 31{,}200$ J/mole.
   (a) Sketch the phase diagram (T versus X_D).
   (b) Sketch the free-energy versus composition diagram at the eutectic temperature.
   (c) Calculate the eutectic temperature and composition of the eutectic melt assuming that the liquid phase is an ideal solution.

9. Consider a binary system A–B, which has a phase diagram similar to that in Figure 1.9. Assume the liquid phase is an ideal solution and the solid phase is a regular solution with $\Omega_s = 15{,}000$ J/mole.
   (a) Calculate the critical temperature and composition of the miscibility gap and the composition of the phase boundaries at a temperature 100 °C below the critical temperature.

(b) Use free-energy versus composition diagrams to explain why the solidus and liquidus curves both have a minimum.
10. Use the Richardson–Ellingham diagram for oxides to answer the following questions.
   (a) What is the dissociation pressure of Cu_2O at 1,000 °C?
   (b) What is the CO/CO_2 ratio in equilibrium with Co and CoO at 800 °C?
   (c) Will an atmosphere with an H_2/H_2O ratio of $10^5/1$ prevent the oxidation of chromium at 1,000 °C?
   (d) What is the standard entropy change for the reaction
   $$2NiO(s) + Si(s) = 2Ni(s) + SiO_2(s)$$
   at 900 °C?
   (e) What is the standard enthalpy change for the reaction in (d)?

Numerous additional problems on bulk thermodynamics are provided at the ends of the chapters in References [1–4].

1.8 Selected thermodynamic data references

[1] *JANAF Thermochemical Tables, J. Phys. Chem. Ref. Data*, Vol. 14, Suppl. 1 (1985).
[2] *JANAF Thermochemical Data*, including supplements (Midland, MI: Dow Chemical Co., 1971); also NSRDS-NBS 37 (Washington D.C.: U.S. Government Printing Office, 1971); supplements 1974, 1975 and 1978.
[3] *JANAF Thermochemical Tables* (3rd edn.) (Midland, MI: Dow Chemical Co., 1986).
[4] Schick, H., *Thermodynamics of Certain Refractory Compounds*, 2 volumes (New York: Academic Press, 1966).
[5] Wicks, C. W. and Block, F. E., *Thermodynamic properties of 65 elements, their oxides, halides, carbides, and nitrides*, Bureau of Mines, Bulletin 605 (Washington D.C.: U.S. Government Printing Office, 1963).

[6] Hultgren, R., Orr, R. L., Anderson, P. D. and Kelley, K. K., *Selected Values of Thermodynamic Properties of Metals and Alloys* (New York: Wiley, 1967).

[7] Kubaschewski, O. and Alcock, C. B., *Metallurgical Thermochemistry*, 5th edn. (Oxford: Pergamon Press, 1979).

[8] Wagman, D. D., and others, *Selected Values of Chemical Thermodynamic Properties, Elements 1 through 34*, NBS Technical Note 270–3 (Washington, D.C.: U.S. Government Printing Office, 1968); and *Elements 35 through 53*, NBS Technical Note 270–4 (Washington, D.C.: U.S. Government Printing Office, 1969).

[9] Stern, K. H. and Weise, E. A., *High Temperature Properties and Decomposition of Inorganic Salts. Part 1. Sulfates*. National Standard Reference Data Series, National Bureau of Standards, no. 7 (Washington D.C.: U.S. Government Printing Office, 1966); and *Part II. Carbonates* (Washington, D.C.: U.S. Government Printing Office, 1969).

[10] Stull, D. R. and Sinke, G. C., *Thermodynamic Properties of the Elements* (Washington D.C.: American Chemical Society, 1956).

[11] Mills, K. C., *Thermodynamic Data for Inorganic Sulphides, Selenides, and Tellurides* (London: Butterworth, 1974).

[12] Dayhoff, M. O., Lippincott, E. R., Eck, R. V. and Nagarajan, G., *Thermodynamic Equilibrium in Prebiological Atmospheres of C, H, O, P, S, and Cl*, NASA SP-3040 (NASA, 1964).

[13] Barin, I. and Knache, O., *Thermochemical Properties of Inorganic Substances* (Berlin: Springer-Verlag, 1973); also supplement, 1977.

[14] Barin, I., *Thermochemical Data of Pure Substances* (Weinheim: VCH Verlagsgesellschaft, 1993).

[15] The *NBS Tables of Chemical Thermodynamic Properties* (Washington, D.C.: National Bureau of Standards, 1982).

[16] Pankratz, J. B., *Thermodynamic Properties of Elements and Oxides*, Bureau of Mines Bulletin 672 (Washington, D.C.: U.S. Government Printing Office, 1982).

[17] Pankratz, J. B., *Thermodynamic Properties of Halides*. Bureau of Mines Bulletin 674 (Washington, D.C.: U. S. Government Printing Office, 1984).

2 Introduction to surface quantities

Since atoms or molecules in the vicinity of the surface of a condensed phase have different bonding from those in the bulk they have different thermodynamic properties. In this chapter the concept of the *Gibbs dividing surface* and the two fundamental quantities for describing the thermodynamic properties of surfaces and interfaces, *surface energy* (γ) and *surface stress* (σ), are defined. The relations between γ and the other thermodynamic variables for surfaces are established. Finally, methods for obtaining γ and σ are described and representative values of both are presented.

2.1 Description of a surface/interface

The atomic/molecular configurations at surfaces and interfaces are different from those in the bulk of a condensed phase. In some instances the changes occur over very short distances, i.e. the interface is "sharp", and in others the differences extend over considerable distances, i.e. the interface is "diffuse". Although considerable progress has been made in both calculating and observing the atomic arrangements at surfaces and interfaces, in the majority of cases, this structure is not known. Therefore, it is necessary to develop models of the surface/interface to allow description of its contribution to the thermodynamic properties of the system.

Two approaches are used to describe the surface/interface. The first is to consider the region of structural variation as a *surface phase* and treat its thermodynamic properties in the same manner as any bulk phase.

The boundaries of the surface phase are usually chosen as locations in the adjoining bulk phases where the structure and local properties are no longer varying with position. The second approach, introduced by Gibbs [1], involves the assumption that all property changes occur at a single plane, the *Gibbs dividing surface*. All of the properties of the two phases are presumed to have their bulk values right to the dividing surface. Therefore, any extensive property of the system may be written according to

$$Q'_{tot} = Q'_\alpha + Q'_\beta + Q^S \qquad (2.1)$$

where Q'_α is the contribution to Q'_{tot} from the α phase, Q'_β is the contribution from the β phase and Q^S is the contribution from the surface. Note that Q^S can be positive, negative or zero for a given system. If Q' is allowed to represent numbers of moles of component A then Equation (2.1) becomes

$$n_A^{tot} = n_A^\alpha + n_A^\beta + n_A^S \qquad (2.2)$$

This approach is illustrated in Figure 2.1 for a hypothetical system A–B, which has two phases α and β in contact. C_A and C_B are the concentrations of A and B, respectively, in units of moles/m^3. Figure 2.1(a) shows the concentration profile of the major component A as a function of distance. Equation (2.2) may be used to express the number of moles at the surface as

$$n_A^S = n_A^{tot} - C_A^\alpha V^\alpha - C_A^\beta V^\beta \qquad (2.3)$$

(Note that, if the profile is defined with unit area normal to the plane of the figure, then $C_A^\alpha V^\alpha + C_A^\beta V^\beta = C_A^\alpha x^\alpha + C_A^\beta x^\beta$.) The position of the dividing surface is generally chosen such that $n_A^S = 0$. That is, the terms $C_A^\alpha V^\alpha + C_A^\beta V^\beta$ account for all the A in the system, n_A^{tot}, and the two "triangular" areas are equal. This choice of position generally means that $n_B^S \neq 0$. This is the case in Figure 2.1(b), where the actual amount

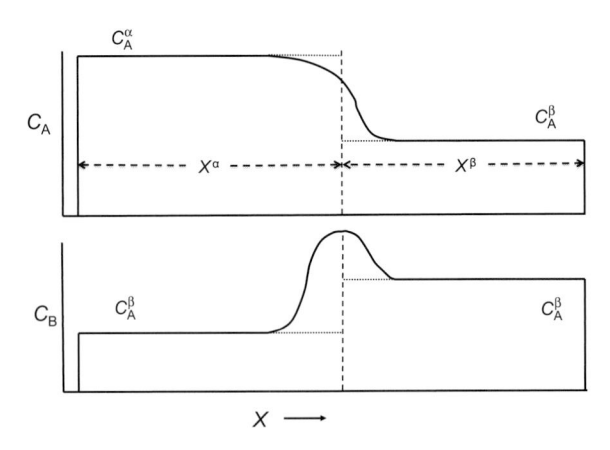

Figure 2.1 The concentration profiles of the major component A and minor component B as a function of distance for a hypothetical system A–B that has two phases α and β in contact. C_A and C_B are the concentrations of A and B, respectively, in units of moles/m³.

of B exceeds that represented by the areas under the rectangular profiles defined by the dividing surface. This means that there is an excess of B at the interface, i.e. it is *adsorbed* ($n_B^S > 0$). It is clear that, for different concentration profiles, n_B^S could be positive, negative or zero. The phenomena of *adsorption* and *desorption* will be described in detail in Chapter 7.

In describing the properties of surfaces it is generally useful to normalize them to unit area of surface. Thus the surface concentrations of A and B are defined as

$$\Gamma_A \equiv \frac{n_A^S}{A} \qquad (2.4)$$

and

$$\Gamma_B \equiv \frac{n_B^S}{A} \qquad (2.5)$$

where A is the area of the interface. These quantities are also called *surface excesses*.

The dividing surface can also be used to define the thermodynamic properties of the surface/interface. For example, if Q' is allowed to represent the entropy of the system then Equation (2.1) becomes

$$S' = S^\alpha + S^\beta + S^S \qquad (2.6)$$

with the three terms on the right-hand side representing the contributions of the α phase, β phase and surface, respectively, to the extensive entropy, S':

$$S^S = S' - S^\alpha_{vol} V^\alpha - S^\beta_{vol} V^\beta \qquad (2.7)$$

The total entropy contribution of the surface is obtained by rearrangement of Equation (2.6) and insertion of the entropies per unit volume of α and β. Thus, S^S is determined by the position of the dividing surface. Finally, normalizing by the surface area yields the surface entropy, s^S:

$$s^S = \frac{S^S}{A} \qquad (2.8)$$

The other thermodynamic properties of the surface or interface may be treated in the same manner.

2.2 Thermodynamic properties

Since atoms or molecules in the vicinity of the surface of a condensed phase have different bonding, they have different thermodynamic properties. Consider the schematic diagram of an hcp crystal in Figure 2.2. The segment of the basal plane in this drawing consist of atoms 1 to 6 plus the shaded atom. In the bulk, the shaded atom would have nearest-neighbor atoms 1 to 12. However, in forming the surface atoms 10, 11 and 12 are removed. Since the atoms above the plane are missing, there is a net force on the surface atoms, which tends to change the extent of the surface and gives rise to a *surface stress*. The surface stress is positive (tensile) if it tends to decrease the extent of the surface, which

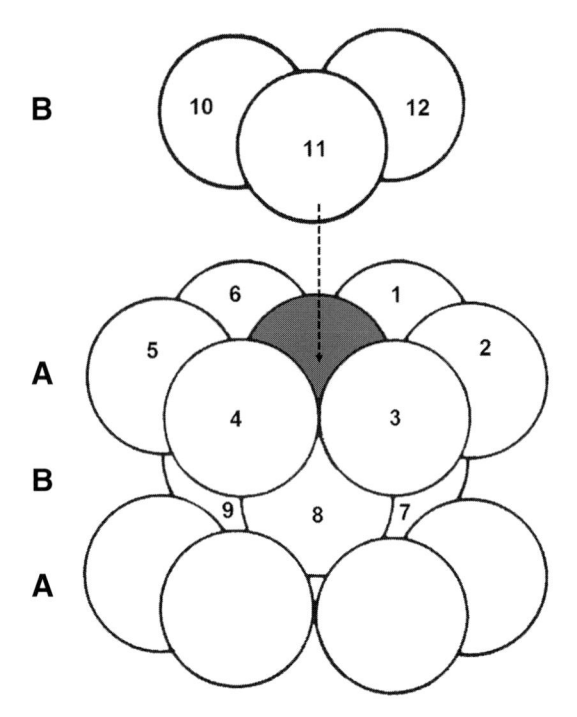

Figure 2.2 A schematic diagram showing the formation of a surface on the basal plane of an hcp crystal.

is commonly the case. Also, since the bonds between the atoms are what stabilize the crystal, the absence of nearest neighbors will result in the surface atoms having higher values of the "energy functions" than atoms in the bulk. This will give rise to properties such as *surface energy*, which describes the excess energy associated with the surface.

Various terms have been used to describe the properties of surfaces. Different authors have used the same name for different properties, e.g. *surface tension*. The following is a set of definitions consistent with the way the thermodynamic functions were defined in Chapter 1. It largely follows the treatments given in the excellent article by Mullins [2] and the monograph by Blakely [3], although the terminology differs slightly, e.g. the term *surface tension* is not used.

2.2.1 Creation of a surface

The fundamental quantity related to the creation of a surface is called the *surface energy* (γ). The quantity γ is defined as the reversible work involved in creating unit area of new surface at constant temperature, volume and total number of moles (J/m^2):

$$\gamma \equiv \lim_{dA \to 0} \frac{\delta w'}{dA} \tag{2.9}$$

The surface energy is a scalar quantity, which will be isotropic for liquids but may be a function of crystallographic orientation for surfaces involving crystalline solids, i.e. it will be a function of the surface plane normal. The latter point can be qualitatively understood by consideration of Figure 2.3 for the case of an fcc crystal. Clearly, a different number of nearest-neighbor atoms will need to be removed to create unit area of a (111) surface as opposed to a (100) surface and, in turn, a (110) surface. Therefore, the energy increase associated with surfaces with each of these orientations will differ. (See Problem 2.1.) Surfaces and interfaces in crystalline systems will be discussed in detail in Chapters 4 and 5.

2.2.2 Extension of a surface

The fundamental quantity associated with extending an existing surface is called the *surface stress* (σ). The quantity σ is defined as the work involved in *reversibly* deforming a surface (N/m).

The term *reversibly* is used here as opposed to the term *elastically*, which is used when the term is restricted to a description of solid surfaces [4]. The surface stress may also be applied to liquids, in which case *elastically* would have no particular meaning. In the case of liquids, σ will be isotropic and will be shown to be numerically equal to γ for unary systems.

46 INTRODUCTION TO SURFACE QUANTITIES

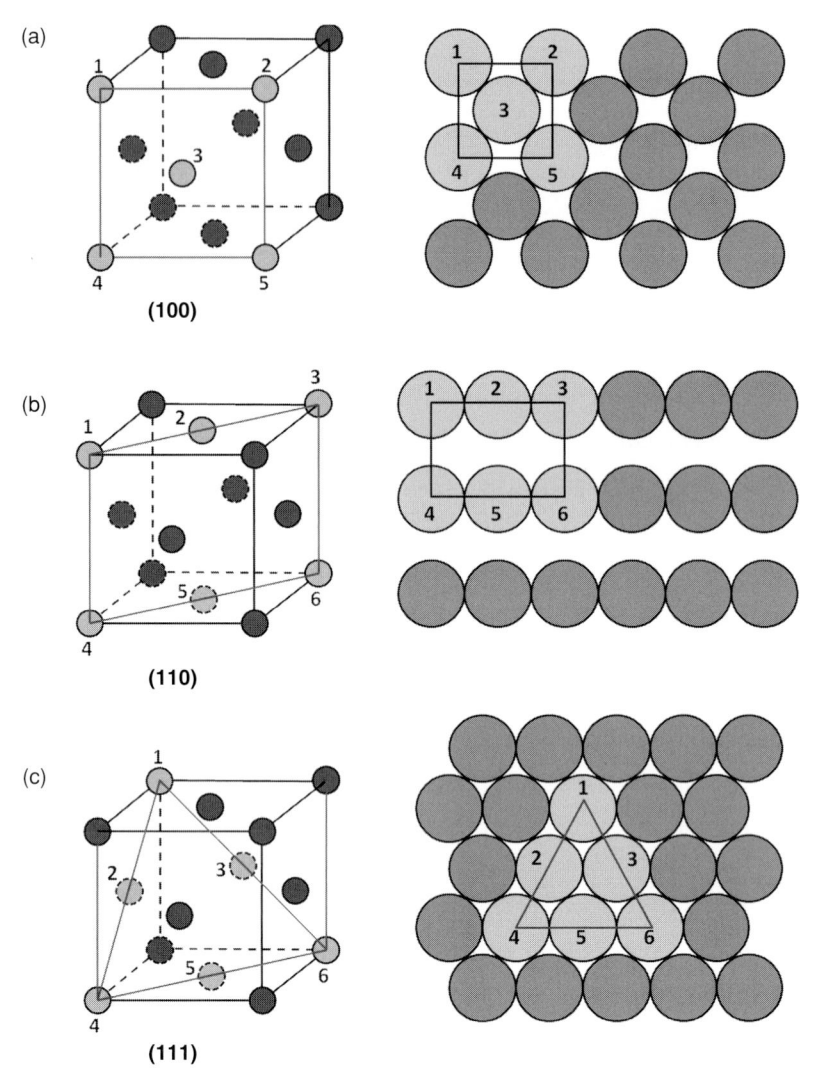

Figure 2.3 A schematic diagram of an fcc crystal showing the atomic packing (a) for a (100) orientation, (b) for a (110) orientation and (c) for a (111) orientation.

Consideration of Figure 2.3 indicates that the surface stress for a crystalline solid will be a function of both the normal of the plane that is being deformed and the direction in which the deformation is taking place. In this case the surface stress will be a second-rank tensor. The surface stress is positive (tensile) if it tends to decrease the extent of the surface, which is always the case for liquids and usually the case for simple solids. However, in principle, the sign of the surface stress can be negative.

2.2.3 Relations among surface quantities

From the definition of γ and Equation (1.32) at constant T and V, the change in Helmholtz free energy for the creation of an increment of planar interfacial area dA for a unary two-phase system will be equal to the work done on the system,

$$dF_{\text{system}} = \delta w' = \gamma \, dA \qquad (2.10)$$

Letting Q' in Equation (2.1) represent the Helmholtz free energy of the system yields

$$dF_{\text{system}} = d(F^\alpha + F^\beta) + dF^S = d(n^\alpha \mu^\alpha + n^\beta \mu^\beta) + dF^S \qquad (2.11)$$

However, since the two phases are equilibrated, the chemical potentials are equal for a planar surface:

$$\mu^\alpha = \mu^\beta = \mu \qquad (2.12)$$

Therefore, Equations (2.10) and (2.11) may be combined to yield

$$\gamma \, dA = \mu d(n^\alpha + n^\beta) + dF^S = \mu \, d(n^{\text{tot}} - n^S) + dF^S \qquad (2.13)$$

The latter term in brackets follows directly from Equation (2.2) and, since n^{tot} is constant, Equation (2.13) reduces to

$$\gamma \, dA = -\mu \, dn^S + dF^S \qquad (2.14)$$

and use of Equation (2.4) results in

$$\gamma\, dA = -\Gamma\mu\, dA + f^S\, dA \tag{2.15}$$

where f^S is defined by

$$f^S \equiv \frac{F^S}{A} \tag{2.16}$$

Thus,

$$\gamma = f^S - \Gamma\mu \tag{2.17}$$

In general, the surface energy and the specific surface Helmholtz free energy are not equal since $\Gamma\mu$ depends on the position of the dividing surface. However, if the dividing surface may be chosen such that $\Gamma = 0$ then

$$\gamma = f^S \tag{2.18}$$

Further relations among the specific surface properties may be developed from the definitions of the various thermodynamic functions. For example, taking $F = F(V, T, A)$ gives

$$dF = -P\, dV - S\, dT + f^S\, dA \tag{2.19}$$

Adding

$$d(PV) = P\, dV + V\, dP \tag{2.20}$$

yields

$$d(F + PV) = dG = V\, dP - S\, dT + f^S\, dA \tag{2.21}$$

Thus, taking $G = G(P, T, A)$ and writing the total differential indicates that the final term in Equation (2.21) corresponds to

$$\left(\frac{\partial G}{\partial A}\right)_{T,P} dA$$

Thus,
$$g^S = f^S \tag{2.22}$$

Again, if a unary system is considered then
$$\gamma = g^S \tag{2.23}$$

Application of the *Maxwell reciprocal relations* to Equation (2.19) yields
$$\left(\frac{\partial f^S}{\partial T}\right)_{V,A} = -\left(\frac{\partial S}{\partial A}\right)_{V,T} \tag{2.24}$$

or, if Equation (2.18) holds,
$$\left(\frac{\partial \gamma}{\partial T}\right)_{V,A} = -s^S \tag{2.25}$$

Equation (2.25) shows that the temperature dependence of the surface energy is determined by the surface entropy, which should generally be positive. Thus, the surface energy decreases as the temperature is increased. From the definition of the Helmholtz free energy, Equation (1.28), the surface Helmholtz free energy, internal energy and entropy are related by
$$f^S = e^S - T s^S \tag{2.26}$$

and, for a one-component system, insertion of Equations (2.18) and (2.25) into Equation (2.26) yields
$$e^S = \gamma - T\left(\frac{\partial \gamma}{\partial T}\right)_V \tag{2.27}$$

Similarly, from the definition of the Gibbs free energy, Equation (1.30),
$$g^S = h^S - T s^S \tag{2.28}$$

and, for a one-component system,
$$h^S = \gamma - T\left(\frac{\partial \gamma}{\partial T}\right)_V \tag{2.29}$$

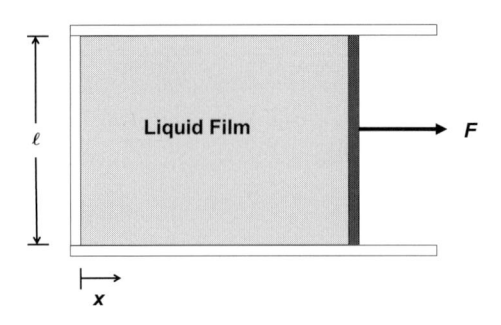

Figure 2.4 A schematic diagram of a liquid film being stretched across a frame and then extended.

where h^S is the *surface enthalpy*. Comparison of Equations (2.27) and (2.29) indicates that

$$h^S = e^S \tag{2.30}$$

The equality in Equation (2.30) results from the fact that $V^S = 0$.

2.2.4 Relations between γ and σ

2.2.4.1 Fluid systems

Consider Figure 2.4, which represents a system consisting of a liquid film, e.g. a soap film, of width l, which is stretched across a frame, the right-hand side (shaded) of which is able to slide. If temperature and pressure are held fixed and a force is applied to the slide wire such that it reversibly moves to the right by a distance dy the work done against the surface stress (work done on the system) will be

$$\delta w'_{T,P} = F\, dx = 2\sigma l\, dx = \sigma\, dA \tag{2.31}$$

The work done by the system will be of the opposite sign. Since this constitutes non-PV work at constant temperature and pressure, from Equation (1.33), the Gibbs free-energy change for the liquid will be

$$dG' = \sigma\, dA \tag{2.32}$$

Considering now that the Gibbs free energy of the liquid may be expressed as $G' = G'(T, P, A)$, the free-energy change will be

$$dG' = \left(\frac{\partial G'}{\partial P}\right)_{T,A} dP + \left(\frac{\partial G'}{\partial T}\right)_{P,A} dT + \left(\frac{\partial G'}{\partial A}\right)_{T,P} dA \xrightarrow{T,P} \gamma \, dA$$

(2.33)

Comparison of Equations (2.32) and (2.33) indicates that

$$\sigma = \gamma \qquad (2.34)$$

Thus for fluid systems the *surface stress* and *surface energy* are numerically identical.

2.2.4.2 Solid systems

The relation between σ and γ for solids can be demonstrated by a concept introduced by Shuttleworth [5] and amplified by Mullins [2] and Cammarata [6]. Consider the single-crystal cube in Figure 2.5 which is taken along two reversible paths.

Path I is as follows. In step 1, the specimen is cut in half along a plane normal to side x. In step 2, both halves are deformed in the y direction under the constraint that the z edge maintains a constant length (the x edge will change) such that area A_0 becomes A. For path I

$$w_\mathrm{I} = w_1 + w_2 = 2A_0\gamma_x + w_{\text{bulk def}} + 4\sigma_{yy} z \, dy \qquad (2.35)$$

where γ_x is the surface energy of the surface with its normal parallel to x and the length z in the final term of Equation (2.35) is equivalent to the length l in the deformation of the liquid in Figure 2.3.

Path II is as follows. In step 3, the specimen is deformed in the y direction. In step 4, the specimen is cut in half along a plane normal to side x. For path II

$$w_\mathrm{II} = w_3 + w_4 = w_{\text{bulk def}} + 2\sigma_{yy} z \, dy + 2(A_0 + z \, dy)(\gamma_x + d\gamma_x)$$

(2.36)

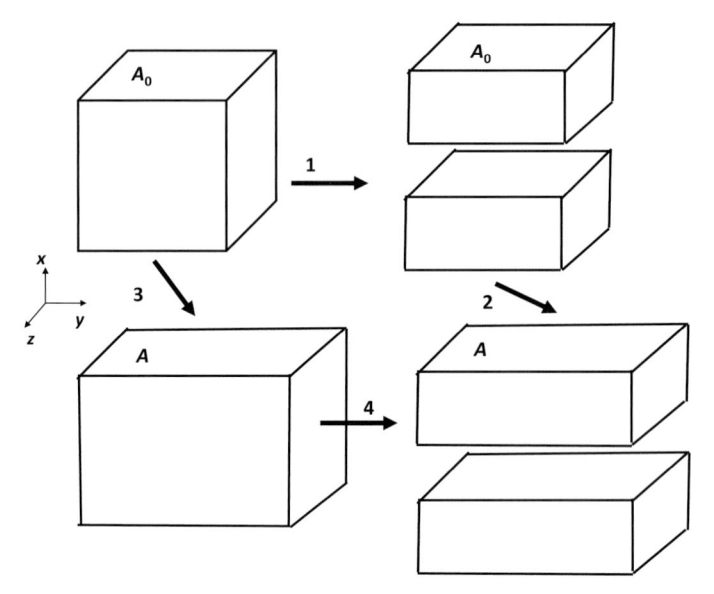

Figure 2.5 A schematic diagram of the application of two equivalent reversible processes involving a single-crystal cube to show the relation between the surface energy and the surface stress [2]. (Reprinted with permission of ASM International.)

Since both paths reversibly bring the crystal to the same state, $w_{\mathrm{I}} = w_{\mathrm{II}}$. Also, the work of bulk deformation will be the same for both paths, so this term subtracts out. Therefore,

$$2A_0 \gamma_x + 4\sigma_{yy} z\, dy = 2\sigma_{yy} z\, dy + 2A_0\, \gamma_x + 2z\, dy\, \gamma_x$$
$$+ 2A_0\, d\gamma_x + 2z\, dy\, d\gamma_x \quad (2.37)$$

The final term is small and may be neglected. Thus, Equation (2.37) simplifies to

$$2\sigma_{yy} z\, dy = 2z\, dy\, \gamma_x + A_0\, d\gamma_x \quad (2.38)$$

or

$$\sigma_{yy} = \gamma_x + \frac{A_0\, d\gamma_x}{z\, dy} \quad (2.39)$$

If the areal strain $d\varepsilon_{zz}$ is defined as

$$d\varepsilon_{zz} = \frac{z\,dy}{A_0} \qquad (2.40)$$

Equation (2.39) becomes

$$\sigma_{yy} = \gamma_x + \frac{d\gamma_x}{d\varepsilon_{zz}} \qquad (2.41)$$

Similarly, for an equivalent process carried out in the z direction at constant y we have

$$\sigma_{zz} = \gamma_x + \frac{d\gamma_x}{d\varepsilon_{zz}} \qquad (2.42)$$

For surfaces that are crystallographic planes with greater than three-fold rotational symmetry σ is isotropic [5] and the relation between surface stress and surface energy may be written simply as

$$\sigma = \gamma + A\frac{d\gamma}{dA} \qquad (2.43)$$

2.2.4.3 Comments on surface stress and strain

As described above, the surface energy and surface stress may be expressed in the same units but are fundamentally different quantities. The surface energy is a scalar quantity and is always positive, whereas the surface stress is a tensor quantity and may be either positive or negative. Clearly the two are numerically equal for liquids and are positive. The establishment of the equality depends on the atoms (molecules) in a liquid being mobile enough to re-establish the original surface structure following deformation. It is evident that this is not the case for solids and the two quantities generally are unequal. The special case of glass-forming systems, such as SiO_2, introduces an interesting nuance to this discussion. Above its melting temperature (1,720 °C) silica is a liquid, and σ and γ would be expected to be equal. However, as the liquid is cooled its viscosity increases and, if crystallization does not occur, at low temperatures the silica is a rigid glass for which σ and γ would be expected to be unequal. This introduces the question as to when the transition occurs. The apparent answer requires the use of the concept of metastability, which was discussed in Section 1.3.2.3. Thus, if the silica is at a temperature at which the surface rearrangement is rapid relative to the time of the measurement it will behave as a liquid, whereas if it is at a temperature at which rearrangement is slow relative to the time of measurement it will behave as a solid. Unfortunately, there are few reliable experimental measurements of surface stress and theoretical values depend strongly on the input assumptions.

2.2.5 Determination of surface parameters

The values of γ and σ for various types of surface are necessary for quantitative application of the thermodynamic relationships developed in Section 2.2.3. These values may be obtained by three approaches.

1. Direct experimental measurement. There are various techniques for measuring γ for surfaces and interfaces. Most of these techniques will be described in subsequent chapters after the phenomena which underlie the measurements have been described. The techniques for measuring σ are rather limited. Selected results of experimentally determined values of γ and σ will be presented in Section 2.2.5.1.
2. Theoretical calculation. Various theoretical methods may be used for calculation of both γ and σ. The results of several techniques will be described in Section 2.2.5.2.
3. Extrapolation using other physical properties. The nature and strength of the bonds between atoms determine not only the values of γ but also a number of other physical properties. Therefore, there is a correlation between γ and these properties. Such correlations are not generally available for the determination of σ.

Table 2.1 *Experimental values of the surface energy (γ) for various types of liquids at the indicated temperature*

Liquid	T (°C)	γ (mJ/m^2)	Reference
H$_2$O	25	72.1	[7]
Ethanol	20	22.4	[7]
Benzene	20	28.9	[7]
Al	600	866	[8]
Al–33 wt% Cu	700	452	[8]
Cu	1,083	1,355	[9]
Au	1,063	1,138	[9]
Fe	1,535	1,855	[9]
Ni	1,455	1,780	[8]
Ni–20 wt% Cu	1,300	1,700	[8]
Ni–40 wt% Cu	1,300	1,460	[8]
Ni–60 wt% Cu	1,300	1,310	[8]
Ni–80 wt% Cu	1,300	1,125	[8]
Ag	951	910	[9]
NaCl	801	114	[10]
Na$_2$SO$_4$	884	196	[10]
Na$_2$SiO$_3$	1,000	250	[10]
Al$_2$O$_3$	2,080	700	[10]

2.2.5.1 Direct experimental measurement of γ and σ

Liquids Since γ and σ are numerically equal for liquids, only the determination of γ is necessary. There are various techniques for measuring γ for liquids. These include the capillary-rise, sessile-drop, pendant-drop, drop-weight and maximum-bubble-pressure techniques. Several of these techniques will be described in Chapters 3 and 6. Table 2.1 presents typical values for the surface energies of various types of liquids.

More extensive tables of γ_{LV} (subscripts L for liquid, V for vapor and S for solid are used throughout) are presented in References [8] and [9]. The following generalizations may be made.

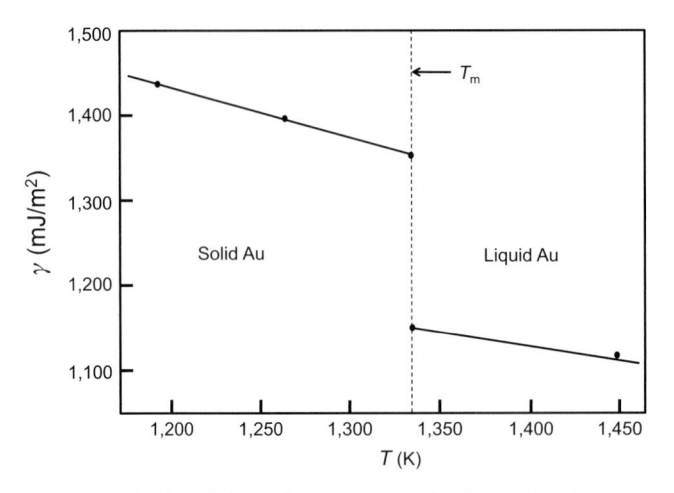

[Figure: A plot showing γ (mJ/m²) vs T (K) for Solid Au and Liquid Au, with a discontinuity at Tm between 1,300 and 1,350 K. Solid Au line decreases from ~1,440 at 1,200 K to ~1,350 at 1,330 K; Liquid Au line decreases from ~1,150 at 1,340 K to ~1,130 at 1,450 K.]

Figure 2.6 A plot of the surface energies of solid and liquid gold versus temperature.

1. The surface energies of organic liquids, e.g. ethanol, are very low.
2. The surface energies of liquid halides, e.g. NaCl, are rather low.
3. The surface energies of liquid metals, e.g. Cu, are quite high.
4. The surface energy of liquids decreases as the temperature increases under the influence of the surface entropy. Reference [9] presents the surface energy of liquid gold as

$$\gamma = 1{,}138 - 0.19(T - T_m)(\text{mJ}/\text{m}^2) \tag{2.44}$$

This expression is plotted in Figure 2.6. The value of -0.19 (mJ/(m² K)) is $d\gamma/dT$ and represents the negative of the surface entropy, s^S, according to Equation (2.25). Examination of the most recent tabulations of γ_{LV} for metals [9] indicates trends for surface entropy depending on the position of the metal in the periodic table. Values of $-d\gamma/dT$ are in the range 0.15–0.2 mJ/(m² K) for the simple metals (Ag, Au, Cu, etc.), 0.4–0.5 mJ/(m² K) for the transition metals, 0.1 mJ/(m² K) for the alkali metals, 0.2–0.25 mJ/(m² K) for the refractory metals and approximately 0.1 mJ/(m² K) for the rare-earth elements.

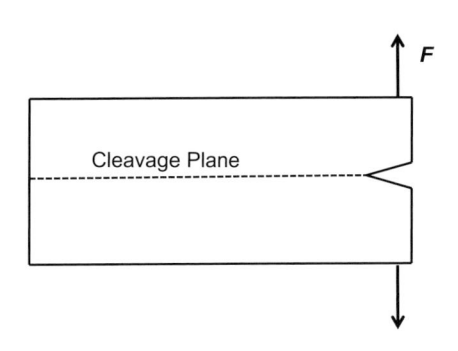

Figure 2.7 A schematic diagram of the cleavage technique for measuring the surface energy of brittle solids.

5. The surface energies of alloys are influenced by composition. This can be seen from Table 2.1 for the case of Ni–Cu alloys, for which γ decreases monotonically with Cu concentration. Similar effects are observed for oxide solutions. For example, additions of Na_2O and P_2O_5 have been shown to dramatically decrease the surface energy of liquid FeO [9].
6. The surface energy of liquid metals can be decreased dramatically by adsorption of *surface-active components*, e.g. O or S, either from the gas phase or as dissolved impurities. (Adsorption phenomena will be discussed in Chapter 7.)

Solids Since γ and σ are generally not equal for solids, both parameters must be measured. The number of techniques for measuring γ for solids is limited and there are very few techniques for measuring σ.

Surface energy The cleavage technique may be used to measure γ for suitable brittle materials, e.g. mica. In this technique the crystal is cleaved by propagating a crack, as shown schematically in Figure 2.7, and the energy of the two new surfaces which are created is equated to the elastic energy of fracture:

$$2\gamma A = \frac{\sigma_{appl}^2 \times A}{2E_{el}} \tag{2.45}$$

where x is the thickness of the specimen, σ_{appl} is the stress caused by the applied force and E_{el} is the elastic modulus of the solid. Equation (2.46) directly yields γ as

$$\gamma = \frac{\sigma_{appl}^2 x}{4E_{el}} \qquad (2.46)$$

This technique is applicable only to brittle ceramics, which may be cleaved with minimal plastic deformation. Since the analysis is based solely on elastic fracture, any plastic deformation will introduce errors into the values determined for γ.

The most common method for measuring γ for solids is the *zero-creep technique*. This technique is based on the observation that thin foils or fine wires will tend to shrink by creep to reduce their surface area and hence the total surface energy. The surface energy is determined by applying an external stress of sufficient magnitude to just prevent the shrinkage. The *multiphase equilibrium technique* [9] may be used for measuring γ for high-melting-point solids, e.g. refractory oxides. This technique uses the effect of a non-reactive liquid on the grain-boundary grooving of polycrystalline materials. The *zero-creep* and *multiphase equilibrium techniques* will be described in Chapter 4. Table 2.2 presents typical values for the surface energies of various types of solids. (Most of these are average values since, generally, the surface energy varies depending on which crystallographic planes make up the surface.)

More extensive tables of γ_{SV} are presented in References [8–10]. The following generalizations may be made.

1. The surface energies of solid halides, e.g. NaCl, are rather low.
2. The surface energies of solid metals, e.g. Cu, are quite high.
3. The surface energy of solid metals is approximately 10% higher than that for the corresponding liquid and decreases as the temperature increases under the influence of the surface entropy. Figure 2.6 includes the temperature-dependent surface energy of

Table 2.2 *Experimental values of the surface energy (γ) for various types of solids at the indicated temperature*

Solid	T (°C)	γ (mJ/m^2)	Reference
Al	450	980	[8]
Al–84 at% Cu	800	1,720	[8]
Cu	970	1,650	[9]
Cu–20 at% Au	850	1,160	[8]
Cu–40 at% Au	850	930	[8]
Cu–60 at% Au	850	910	[8]
Cu–80 at% Au	850	1,140	[8]
Au	1,000	1,400	[8]
Fe (δ phase)	1,450	1,950	[8]
Fe (γ phase)	1,350	2,100	[8]
Ni	1,060	2,280	[8]
Ag	950	1,100	[8]
NaCl (100)	25	300	[10]
LiF (100)	−196	340	[10]
Al$_2$O$_3$ (average)	1,850	905	[10]

solid gold, which was calculated using the value at 1,273 K in Table 2.2 and a value of −0.43 mJ/(m^2 K) for $d\gamma/dT$ [8]. Murr [8] has proposed a general correlation between the surface energies for cubic metals and their liquids as

$$\gamma_{SV} = 1.2\gamma_{LV}^{T_m} + 0.45(T_m - T) \qquad (2.47)$$

This correlation agrees very well with the data for solid Au in Figure 2.6. However, as with the liquid metals, there is a significant variation in the experimental values for $d\gamma_{SV}/dT$ [8], which range from −0.2 to −0.9 mJ/(m^2 K). However, there is an insufficient number of experimental measurements for one to attempt to draw correlations as was done above for $d\gamma_{LV}/dT$.

4. The surface energies of alloys are influenced by composition. This can be seen from Table 2.2 for the case of Cu–Au.
5. The surface energy of solid metals can be decreased dramatically by adsorption of *surface-active components*, e.g. O or S, either from the gas phase or as dissolved impurities. (Chapter 7.)

Surface stress There have been few experimental determinations of surface stress. Cammarata [6] has summarized most of these. The surface stress will induce a radial elastic strain (ε) in small spheres, which can be measured by electron diffraction. This strain has been related to the surface stress using the Laplace equation (see Chapter 6) by

$$\sigma = \tfrac{3}{2} K \varepsilon r \tag{2.48}$$

where r is the radius of the sphere and K is the bulk modulus. This technique yielded values of σ equal to 1,175 mJ/m^2 for Au and 1,415 mJ/m^2 for Ag at temperatures near 50 °C. Using Equation (2.47) one would estimate γ_{SV} for Au to be 2,380 mJ/m^2 at 50 °C. Thus, σ would be about 50% of the extrapolated value of γ_{SV}. The accuracy of the measured value of σ might not be high, given problems relating elastic bulk and surface strains [11], but the sign is correct.

2.2.5.2 Theoretical calculations of γ and σ

Theoretical calculations involve determinations of the changes that occur when a surface is created (γ) or stretched (σ). The details depend on the nature of the bonding in the materials and hence the potential energies of interaction between atoms or ions. Kittel [12] has reviewed the bonding in crystals. The interactions in all cases involve two terms: an attractive term, which brings the atoms (ions) together and extends over significant separation distances; and a short-range repulsive term, which increases rapidly as the inner cores start to overlap. Figure 2.8 is a schematic diagram of how these terms add to give a potential-energy curve with

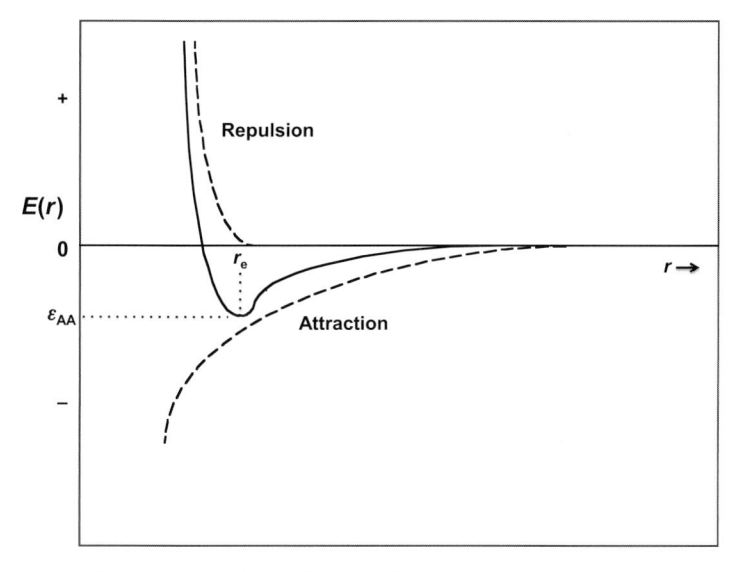

Figure 2.8 A generic plot of the potential energy of interaction versus the atomic (ionic) separation.

a minimum at the equilibrium separation r_e. The detailed shape of this curve will depend on the type of bonding. The types of solids may be summarized as follows.

2.2.5.2.1 Common bonding types

Inert-gas crystals The electron distribution is similar to that for the free atoms. Therefore, the atoms behave similarly to hard spheres, which pack together to form crystals with a close-packed structure, typically fcc. The atoms are bound with the weak van der Waals dispersion force which results from the oscillation of the electrons around the nuclei. The potential energy of interaction between two atoms may be described by the Lennard-Jones potential given by

$$E(r) = -\frac{A}{r^6} + \frac{B}{r^{12}} \tag{2.49}$$

where *A* and *B* are material constants and *r* is the separation distance. The first term describes the attractive interaction and the second term represents the short-range repulsive interaction.

Ionic crystals When the constituent atoms have significant differences in electronegativity, they tend to form ionic compounds, with the electropositive atom contributing electrons and becoming a positively charged cation and the electronegative atom accepting electrons to become a negatively charged anion, e.g. NaCl. The Coulombic attraction between the oppositely charged ions results in them packing into an ionic solid. The potential energy of interaction between two ions may be represented by

$$E(r) = \frac{Z_i Z_j e^2}{4\pi \varepsilon_0 r_{ij}} + \frac{B_{ij}}{r_{ij}^n} \qquad (2.50)$$

where the first term describes the Coulombic interaction (negative for oppositely charged ions and positive for ions with the same sign of their charge). $Z_i e$ and $Z_j e$ are the charges on the two ions, ε_0 is the permittivity of free space and r_{ij} is the separation distance. The second term describes the short-range repulsion. The exponent *n* in the repulsive term typically has values on the order of 10. In some instances the $1/r^n$ repulsive term is replaced with an exponential function of *r*. The total energy is determined by summing the interactions of all the ions. The performance of this summation is described in Chapter 1 of Reference [10].

Covalent crystals Covalent bonds are formed when atoms share electrons. The formation of covalent crystals results in highly directional bonding and packing, which is far from close-packed.

Metallic crystals Metallic bonding results from the lowering of the energy of the conduction electrons as they interact with multiple nuclei. The crystals thus approximate packing of hard spheres. The simplest metals can be fairly well described as Lennard-Jones solids. The bonding in metals may also be described using embedded-atom potentials, in which method the energy is calculated by assuming each atom is embedded in the local electron density contributed by the other atoms [13].

2.2.5.2.2 Calculation methods A number of methods may be used to calculate the energy of crystals using a suitable potential. The difference between the energy of the perfect crystal and that of the crystal containing defects (e.g. vacancies, surfaces, etc.) is generally used to determine the defect energy. The more commonly used techniques for calculation of surface parameters have been summarized by Howe [13].

Monte Carlo An initial configuration of atoms with known interaction potentials is placed in a finite box with fixed boundary conditions at fixed temperature. Individual atoms are moved randomly and if the energy decreases the new arrangement is accepted. If the energy increases the new arrangement is rejected, or kept with a weighting probability that decreases with increasing energy. Calculations are repeated until the configuration that gives the minimum energy is achieved.

Molecular dynamics The initial positions for the atoms are chosen at a particular temperature and a particular interaction potential is assumed. The equations of motion are solved to follow ensuing transitions until constant thermodynamic properties are achieved. This allows determination of the atom positions (structure), energies (surface energy) and forces on the atoms (surface stress) observed at any point in the simulation.

2.2.5.2.3 Selected theoretical values of γ and σ Broughton and Gilmer [14] used molecular dynamics to calculate the surface energy and surface stress for a unary system with a P–T diagram of the type in Figure 1.4 for which the atoms were assumed to interact according to a Lennard-Jones potential. The solid phase was assumed to take the fcc structure. The results from this simulation for temperatures and pressures near

Table 2.3 *Calculated values of the surface energy (γ) and surface stress (σ) for solid/vapor, liquid/vapor and solid/liquid interfaces for a hypothetical fcc Lennard-Jones crystal [14]*

Surface	Surface normal	γ (mJ/m^2)	σ (mJ/m^2)
Solid/vapor	[111]	1,170	0
	[100]	1,150	700
	[110]	1,170	−100
Solid/liquid	[111]	350	−800
	[100]	340	0
	[110]	360	−700
Liquid/vapor		750	750

Table 2.4 *Calculated values of the surface energy (γ) and surface stress (σ) for (111) surfaces of several fcc metals*

Metal	γ (mJ/m^2)	σ (mJ/m^2)
Al	960	1,250
Pt	2,190	5,600
Au	1,250	2,770
Pb	500	820

the triple point are presented in Table 2.3. The values for the surface energies seem reasonable and might approximate a simple metal such as copper, silver or gold. However, the values for the surface stress are rather counter-intuitive, particularly the great variability and large negative values for the solid/liquid interface.

Table 2.4 compares values of γ and σ for (111) surfaces of several simple fcc metals, which were calculated using a pseudopotential approach [6]. All the values of σ are positive (i.e. tensile), which is reasonable considering the simple structure of the metals.

Tasker [15] calculated the surface energies and surface stresses for a group of alkali halide crystals using empirical potentials similar to

Table 2.5 *Calculated values of the surface energy (γ) and surface stress (σ) for (111) surfaces of several alkali halides [15]*

Crystal	Surface normal	γ (mJ/m^2)	σ (mJ/m^2)
LiF	[100]	416	2,533
LiCl	[100]	177	495
NaF	[100]	321	1,056
NaCl	[100]	180	502
LiF	[110]	975	1,048
LiCl	[110]	420	134
NaF	[110]	717	501
NaCl	[110]	392	275

Equation (2.51). Selected results from these calculations are presented in Table 2.5. (Only one value was calculated for the (110) orientation.) The surface energies are of similar magnitude to the available experimental values (Table 2.2) and show a fairly significant orientation dependence. The surface stresses are all positive, which again is consistent with the simple (NaCl-type) crystal structure.

The above calculations suggest that for simple crystal structures in metals and ionic crystals the surface stresses are positive. This is consistent with a qualitative consideration of how the surface atoms (ions) interact with the underlying atoms in the crystal, i.e. the net attractive interactions should be stronger than the net repulsive interactions. However, for more complex structures σ can be zero or even negative for certain orientations. For example, III–IV semiconducting compounds, such as GaAs, can form two types of (111) surfaces. One surface is terminated by a plane of Ga atoms and the other is terminated by a plane of As atoms [6]. The calculated value for the former surface is $-1,000$ mJ/m^2 and for the latter is $+500$ mJ/m^2. Unfortunately, the lack of experimental measurements of σ makes evaluation of the accuracy of the calculations difficult. Finally, it should be re-emphasized that, although the units of γ and σ are equivalent they are two different quantities (γ is a scalar and σ is a second-rank tensor), which describe two different processes.

2.2.5.3 Correlations with other physical properties

The surface energies of solids and liquids are determined by the bonding in the bulk materials and therefore one would expect the surface energy to scale with other physical properties. Consider the simple bond-breaking approach that was used in Section 1.2.2 to develop the regular solution model. Consider the creation of unit-area surface of a liquid. The change in energy will be determined by breaking the bonds across the surface plane. Therefore,

$$\gamma_{LV} = -N_{Bonds}\frac{\varepsilon_{AA}}{2} \qquad (2.51)$$

where N_{Bonds} is the number of bonds broken in forming the surface. This will be related to the area occupied by each atom and the number of bonds broken, and may be expressed as

$$N_{Bonds} = \frac{Zf_{AA}}{\omega} \qquad (2.52)$$

where f_{AA} is the fraction of the bonds that are broken and ω is the area of the surface occupied by an atom. Substitution of Equation (2.52) along with Equation (1.58) for ε_{AA} into Equation (2.51) yields

$$\gamma_{LV} = \frac{f_{AA}\Delta H_A^{vap}}{\omega N_0} \qquad (2.53)$$

Since ω will be proportional to the molar volume of the liquid to the power 2/3, Equation (2.53) suggests a relationship of the type

$$\gamma_{LV} = C\frac{\Delta H_A^{vap}}{V^{2/3}} \qquad (2.54)$$

The surface energies of a large number of liquid metals are observed to follow this correlation, with $C = 0.16 \times 10^{-8}$ when ΔH^{vap} is expressed in mJ [9]. A linear dependence on $\Delta H_A^{vap}/V^{2/3}$ is also found for liquid halides. Since the entropy of vaporization, $\Delta H_A^{vap}/T_b^A$, for metals is approximately constant according to Trouton's rule [16] at 88 J/(mole K) for $T_b < 4{,}000$ K the surface energies of liquid metals also scale with the boiling point. A less obvious correlation is that with the melting point of the metal [8].

Using the bond-breaking approach to create a solid surface would yield

$$\gamma_{SV} = \frac{f_{AA} \Delta H_A^{subl}}{\omega N_0} \tag{2.55}$$

From this expression the orientation dependence of γ_{SV} may also be estimated, since f_{AA} will vary depending on the orientation of the surface plane. (See Problem 1.) Use of an averaged value for f_{AA} in Equation (2.55) would suggest a correlation for γ_{SV} of the type

$$\gamma_{SV} = C' \frac{\Delta H_A^{subl}}{V^{2/3}} \tag{2.56}$$

Indeed the surface energies for many solid metals are found to correlate with $\Delta H_A^{subl}/V^{2/3}$.

2.2.6 Description of surface contributions to the thermodynamic description of material systems

Depending on the experimental details and constraints, the proper surface quantities to appropriately describe different material systems differ.

Fluid systems In the case for which all the phases are fluids $\sigma = \gamma$ for each surface and interface involved and there is generally no confusion.

Solid systems The description of the thermodynamics of systems involving solids becomes more complicated when surface quantities must be

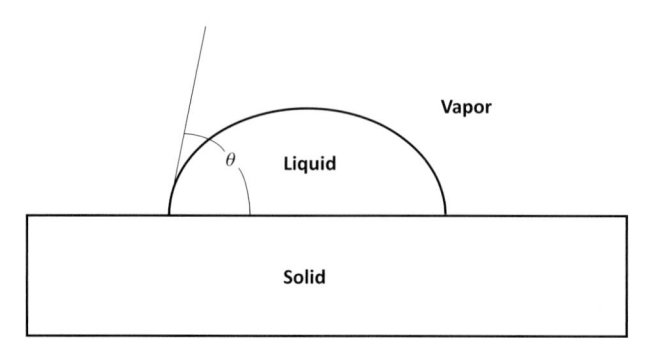

Figure 2.9 A schematic diagram of a liquid droplet only partially wetting the solid surface on which it rests.

considered. These differences were first treated by Shuttleworth [5] and subsequently were described in more detail by Cahn [4].

1. Single-component liquid lying on a solid. The tendency for a liquid to spread across a solid (i.e. to wet the solid) is described by the wetting angle θ_Y as illustrated in Figure 2.9. The wetting angle may be calculated from the Young equation:

$$\cos\theta_Y = \frac{\gamma_{SV} - \gamma_{SL}}{\gamma_{LV}} \qquad (2.57)$$

Thus, θ_Y is determined by a balance of the surface energy and γ is the proper function to use. A simple force balance about the triple point using σ_{SV}, σ_{SL} and σ_{LV} will not yield the correct value for θ because the complete force balance includes a stress in the solid since, in general, $\sigma_{SV} \neq \gamma_{SV}$ and $\sigma_{SL} \neq \gamma_{SL}$. This will be discussed in detail in Chapter 3.

2. Vapor pressure of a liquid droplet or solid particle. The increase in vapor pressure of a liquid droplet as the droplet radius is decreased is given by the Kelvin equation:

$$RT \ln\left(\frac{p(r)}{p(\infty)}\right) = \frac{2\gamma_{LV} V^\circ}{r} \qquad (2.58)$$

The same equation describes the vapor pressure of an isotropic solid. The description of the vapor pressure of a small crystal becomes a little more complex (see Chapter 6), but the appropriate surface function to describe the equilibrium is still γ.

3. Melting point of a one-component solid. The lowering of the melting point of an isotropic solid as the particle radius is decreased is given by

$$T_m(\infty) - T_m(r) = \frac{2\gamma_{SL} V^\circ}{r(S_L - S_S)} \approx \frac{2\gamma_{SL} T_m(\infty)}{r \Delta H_m} \quad (2.59)$$

4. Pressure inside a gas bubble in an isotropic solid. The excess gas pressure inside a bubble in an isotropic solid is given by the Laplace equation:

$$\Delta P = \sigma_{SV} \left(\frac{1}{r_1} + \frac{1}{r_2} \right) \quad (2.60)$$

which, for the special case of a spherical bubble, is

$$\Delta P = \frac{2\sigma_{SV}}{r} \quad (2.61)$$

The general rule when solid phases of small dimensions are involved is that, if a process involves work against the surface stress, σ is used, whereas, if the process creates new surface area, γ is used. Thus, in expressions for the relations of chemical potentials for multicomponent, multiphase systems both σ and γ may appear. This will be discussed in more detail in Chapter 6.

2.3 Summary

The concept of the *Gibbs dividing surface* and the two fundamental quantities for describing surfaces and interfaces, the *surface energy* and *surface stress*, have been defined in this chapter. The units of γ and σ are equivalent, but they are two fundamentally different quantities (γ is a scalar and σ is a second-rank tensor), which describe two different processes. The two functions are numerically equal for liquids but differ for solids. Experimentally measured surface energies are available for many liquids and some solids. There have been very few measurements of surface stress, and some theoretical calculations result in physically unreasonable values. The relations between γ and the other thermodynamic variables for surfaces have been described. The surface energy and surface stress will be used to treat specific aspects of the thermodynamics of surfaces and interfaces in subsequent chapters.

2.4 References

[1] J. W. Gibbs, *Collected Works*, Vol. 1 (New Haven, CT: Yale University Press, 1957).
[2] W. W. Mullins, Solid surface morphologies governed by capillarity. In *Metal Surfaces: Structures, Energetics and Kinetics*, ed. W. D. Robertson and N. A. Gjostein (Metals Park, OH: ASM, 1963), Chapter 2, pp. 17–66.
[3] J. M. Blakely, *Introduction to the Properties of Crystal Surfaces* (Oxford: Pergamon Press, 1973).
[4] J. W. Cahn, Surface stress and the chemical equilibrium of small crystals – I. The case of isotropic surface, *Acta Metall.*, **28** (1980), 1333–1338.
[5] R. Shuttleworth, The surface tension of solids, *Proc. Phys. Soc. (London)*, **A63** (1950), 444–457.
[6] R. C. Cammarata, Surface and interface stress effects in thin films, *Prog. Surf. Sci.*, **46** (1994), 1–37.
[7] A. W. Adamson and A. P. Gast, *Physical Chemistry of Surfaces*, 6th edn. (New York: John Wiley & Sons, 1997), Chapter 1.
[8] L. E. Murr, *Interfacial Phenomena in Metals and Alloys* (Reading, MA: Addison-Wesley Publishing Co., 1975), Chapter 3.
[9] N. Eustathopoulos, M. G. Nicholas and B. Drevet, *Wettability at High Temperatures* (Kidlington: Elsevier Science, 1999), Chapter 4.
[10] Y.-M. Chiang, D. Birnie III and W. D. Kingery, *Physical Ceramics* (New York: John Wiley & Sons, 1997), Chapter 5.

[11] P. R. Couchman, W. A. Jesser, D. Kuhlmann-Wilsdorf and J. P. Hirth, On the concepts of surface stress and strain, *Surf. Sci.*, **33** (1972), 429–436.

[12] C. Kittel, *Introduction to Solid State Physics*, 6th edn. (New York: John Wiley & Sons, 1986), Chapter 3.

[13] J. M. Howe, *Interfaces in Materials* (New York: John Wiley & Sons, 1997), Chapter 1.

[14] J. Q. Broughton and G. H. Gilmer, Surface free energy and stress of a Lennard-Jones crystal, *Acta Metall.*, **31** (1983), 845–851.

[15] P. W. Tasker, The surface energies, surface tensions and surface structure of the alkali halide crystal, *Phil. Mag. A*, **39** (1979), 119–136

[16] D. R. Gaskell, *Introduction to the Thermodynamics of Materials*, 5th edn. (New York: Taylor and Francis, 2008), Chapter 6.

2.5 Study problems

1. Use the quasichemical model to estimate the following quantities for the surfaces of fcc copper having the three orientations depicted in Figure 2.3:
   (a) the surface energies at 298 K and 1,000 K; and
   (b) the surface entropy (assume this quantity to be independent of temperature).
   The heat of sublimation of Cu is $\Delta H_S = 337.6$ kJ/mole at 298 K and $\Delta H_S = 333.5$ kJ/mole at 1,000 K. The lattice parameter is $a_0 = 0.3615$ nm at 298 K. The linear coefficient of thermal expansion is $\alpha_l = 16.5 \times 10^{-6}$ K^{-1}.

2. Consider the three planes in Figure 2.3. What is the rotational symmetry of each? How many values of the surface stress will be needed to describe the deformation of each surface?

3. Consider the combined first and second laws in terms of the Gibbs free energy, Equation (2.21). How many Maxwell reciprocal relations can be obtained from this equation? Write each of them and comment on their physical significance.

4. Copper and nickel form liquid solutions, which are essentially ideal.
   (a) Consider the dissolution of one mole of Ni into nine moles of liquid Cu, which is held at 1,400 K. Assume that the Ni is present as large chunks and has been preheated to 1,400 K. Calculate the changes in enthalpy and entropy for the system.
   (b) Now consider the dissolution process with the Ni present as 200-nm-diameter spheres. Will the enthalpy and entropy changes be different? If so, by how much will they differ?

   For Ni, $T_m = 1{,}726$ K, $\Delta H_m = 17{,}200$ J/mole, $C_p(\text{solid}) = 29.7 + 4.18 \times 10^{-3}T - 9.33 \times 10^5 T^{-2}$ J/(mole K), $C_p(\text{liquid}) = 43.10$ J/(mole K), $V = 6.60 \times 10^{-6}$ m^3/mole. Take $d\gamma_{SV}/dT = -0.55$ mJ/(m^2 K). Show that the calculations in parts (a) and (b) are consistent with the second law of thermodynamics.

3 Equilibrium at intersections of surfaces: wetting

There are many technologies for which the spreading (wetting) of a liquid on a solid surface is important. These include joining of metals and ceramics by brazing, solidification of liquid metals in ceramic molds and fabrication of metal–matrix composites by infiltration of a liquid metal into a fiber preform of non-metallic material. This chapter will present a brief comparison between "non-reactive wetting" and "reactive wetting", followed by a discussion of the fundamental aspects of wetting, which have been developed for non-reactive systems. It will conclude with a brief description of the complicating factors associated with reactive systems such as those at high temperatures.

3.1 Non-reactive versus reactive wetting

The distinction between non-reactive and reactive wetting is based on the extent of the interaction between the solid and liquid phases. These will be determined by the nature of the bonding between the components of the two phases. When the bonding consists of weak van der Waals-type bonds the interactions will be minimal and the wetting will be *non-reactive*. Generally this will be the case for relatively inert materials near room temperature. When the bonding involves stronger covalent and/or ionic interactions the wetting will be *reactive*. This is often the case at elevated temperatures and can include dissolution of the solid into the liquid and formation of new phases at the solid/liquid interface. Reactive wetting can also include interaction of the liquid

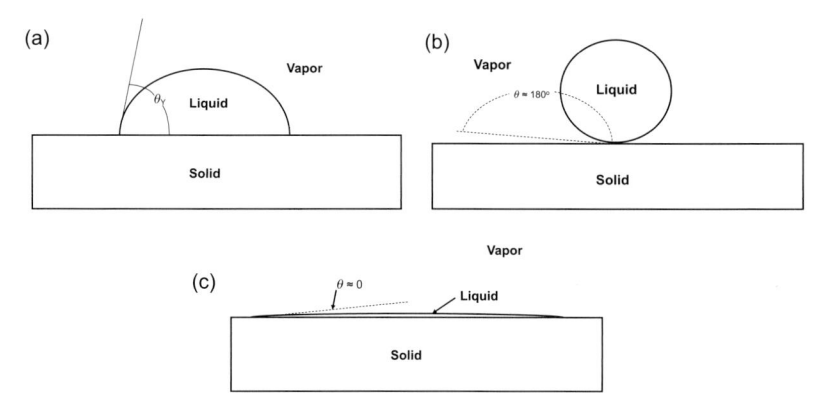

Figure 3.1 A schematic diagram of a liquid droplet that (a) only partially wets the solid surface on which it rests, (b) is completely non-wetting and (c) completely wets the surface.

and/or solid with components from the vapor phase. In these cases the relevant surface energies are often time-dependent.

3.2 Non-reactive wetting

3.2.1 The contact angle on an ideal solid surface (Young's equation)

The wetting of a solid by a liquid was first quantified by Young [1]. The possible configurations for a liquid (*sessile drop*) on a perfectly smooth, non-deformable, non-reactive solid surface are shown schematically in Figure 3.1. If the liquid partially wets the surface it will take a configuration such as that shown in Figure 3.1(a). The extent of wetting is determined by the surface and interfacial free energies between each of the phases, which define the *wetting angle* θ_Y. The equilibrium value of θ may be determined as follows by considering a surface projection of a sessile drop, Figure 3.2. If gravity is neglected and the system is assumed to be at equilibrium at constant temperature and pressure then

$$\frac{dG'}{dr} = 0 \tag{3.1}$$

3.2 NON-REACTIVE WETTING

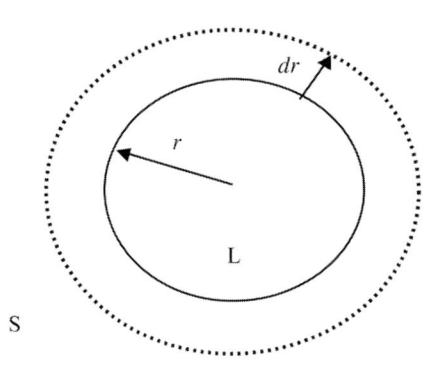

Figure 3.2 A projection of a sessile drop.

An incremental change in the interfacial area, and hence in θ, results in a free-energy change

$$\frac{dG'}{dr} = g_{LV}^S \frac{dA_{LV}}{dr} + g_{SV}^S \frac{dA_{SV}}{dr} + g_{SL}^S \frac{dA_{SL}}{dr} = 0 \quad (3.2)$$

The changes in areas in Equation (3.2) will have the following relationships:

$$\frac{dA_{SL}}{dr} = -\frac{dA_{SV}}{dr} \quad (3.3)$$

and

$$\frac{dA_{LV}}{dr} = \cos\theta \frac{dA_{SL}}{dr} \quad (3.4)$$

so that Equation (3.2) becomes

$$g_{LV}^S \left(\cos\theta \frac{dA_{SL}}{dr}\right) + g_{SV}^S \left(-\frac{dA_{SL}}{dr}\right) + g_{SL}^S \frac{dA_{SL}}{dr} = 0 \quad (3.5)$$

or

$$\cos\theta_Y = \frac{g_{SV}^S - g_{SL}^S}{g_{LV}^S} \quad (3.6)$$

Whenever Equation (2.23) may be applied, one has

$$\cos\theta_Y = \frac{\gamma_{SV} - \gamma_{SL}}{\gamma_{LV}} \quad (3.7)$$

which is known as *Young's equation*. The value of θ_Y will be determined by the relative magnitudes of the various surface energies. Figure 3.1(b) shows the situation when the liquid does not wet the surface at all ($\cos\theta_Y \leq -1$ or $\theta_Y = 180°$), i.e. the liquid *dewets* the solid. Figure 3.1(c) shows the situation where the liquid completely wets the solid ($\cos\theta_Y \geq 1$ or $\theta_Y = 0°$).

3.2.1.1 Comments on Young's equation

Young's equation was derived by minimization of the free energy of the system. Clearly the equilibrium drop configuration must also satisfy a force balance. Examination of Figure 3.1 indicates that, if just the surface stresses at the point of intersection are considered, there would be an apparent imbalance. The forces in the horizontal direction might or might not be balanced, depending on the relative magnitudes of σ_{SV}, σ_{SL} and σ_{LV}. However, the forces in the vertical direction would appear to have an unbalanced component $\sigma_{LV} \sin\theta_Y = \gamma_{LV} \sin\theta_Y$. This vertical component will be balanced by a stress in the solid substrate. At low temperatures this stress will result in elastic deformation of the substrate, but for most solids the magnitude of this deformation will be too small to be observed [2]. Therefore, the Young equation corresponds to a state of metastable equilibrium. Note that at elevated temperatures, if the substrate can creep, it will deform into a ridge under the influence of the surface stress.

The validity of the Young equation has been questioned, in part, because of the above-mentioned metastability and possible effects of gravity [3]. However, Garandet *et al.* [4] have performed an analysis of the sessile-drop problem by minimizing the free energy of the system including gravity. The result indicates that, while gravity can affect the overall shape of the drop, Figure 3.3, the contact angle is still determined by Young's equation.

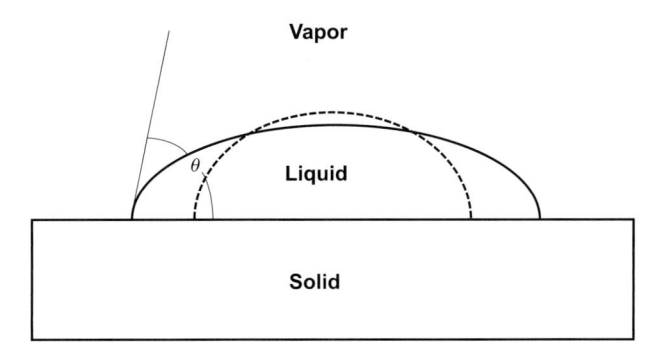

Figure 3.3 A schematic diagram of a sessile drop under the influence of gravity (solid line) and under zero gravity (dashed line).

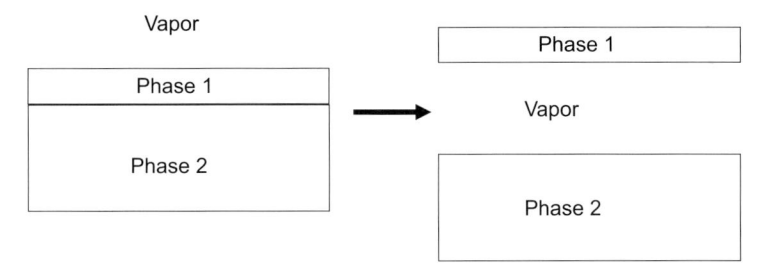

Figure 3.4 A schematic diagram of the separation of two phases to demonstrate the work of adhesion.

3.2.2 Work of adhesion

A quantity related to Young's equation is the *work of adhesion*, the work required to separate two adherent condensed phases, as illustrated in Figure 3.4 This term consists of the energy of the two new free surfaces which are created minus the energy of the interface which is destroyed.

$$W_{ad} = \gamma_{2V} + \gamma_{1V} - \gamma_{12} \tag{3.8}$$

If phase 1 represents liquid and phase 2 represents solid,

$$W_{ad} = \gamma_{SV} + \gamma_{LV} - \gamma_{SL} \tag{3.9}$$

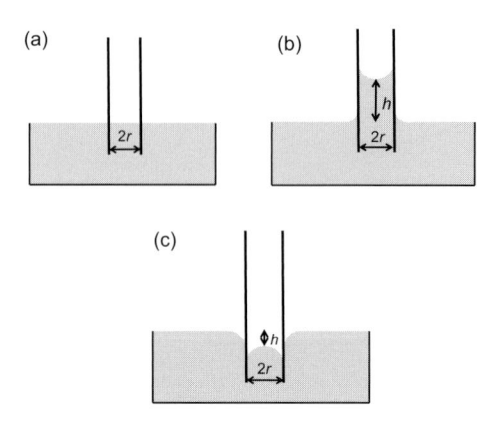

Figure 3.5 A capillary-rise experiment when $\gamma_{SV} - \gamma_{SL}$ is (a) zero, (b) positive and (c) negative.

Substitution of $\gamma_{SV} - \gamma_{SL}$ from Equation (3.7) shows that the work of adhesion is related to the contact angle through the Young–Dupré equation [5],

$$W_{ad} = \gamma_{LV}(1 + \cos\theta_Y) \qquad (3.10)$$

The work of adhesion is an important contributor in the important practical subject of the adherence of coatings to substrates. This will be discussed in detail in Chapter 8.

3.2.3 Capillary rise

A simple application of the concept of wetting is the use of the *capillary-rise* technique to measure the surface energy of a liquid. The experimental arrangement is shown schematically in Figure 3.5(a). If $\gamma_{SV} - \gamma_{SL} = 0$, the liquid will remain in the initial position. If the liquid wets the material of which the capillary is made ($\gamma_{SV} - \gamma_{SL} > 0$), the liquid will rise to a height h (Figure 3.5(b)). The driving force for this rise is the replacement of the solid surface with a liquid/solid interface. The capillary rise may be derived by writing an energy balance for the system. The mass of the liquid in the column is

$$\rho V' = \rho \pi r^2 h \qquad (3.11)$$

where ρ is the density of the liquid and, since the center of gravity is at $h/2$, the increase in potential energy of the liquid will be

$$\text{PE}_\text{L} = \tfrac{1}{2}\rho\pi r^2 h^2 g \qquad (3.12)$$

(Rigorously ρ should be replaced by the difference in density between the liquid and the gas which is displaced, e.g. air, but for most liquids of interest the correction is negligible.) There will also be a decrease in surface energy associated with replacing solid surface with solid/liquid interface,

$$E_\text{Surf} = (\gamma_\text{SV} - \gamma_\text{SL})(2\pi r h) = \gamma_\text{LV}\cos\theta_\text{Y}(2\pi h r) \qquad (3.13)$$

Thus, for a given value of h, the total change in energy will be

$$E_\text{total} = \tfrac{1}{2}\rho\pi r^2 h^2 g - \gamma_\text{LV}\cos\theta_\text{Y}(2\pi h r) \qquad (3.14)$$

The equilibrium value of h will be that which minimizes E or

$$\frac{dE_\text{total}}{dh} = 0 \qquad (3.15)$$

Thus differentiating Equation (3.14) with respect to h and substituting into Equation (3.15) yields upon rearrangement

$$h = \frac{2\gamma_\text{LV}\cos\theta_\text{Y}}{\rho g r} = \frac{2(\gamma_\text{SV} - \gamma_\text{SL})}{\rho g r} \qquad (3.16)$$

If $\gamma_\text{SV} - \gamma_\text{SL} < 0$, the liquid will be depressed in the capillary as shown in Figure 3.5(c).

The quantity $(\gamma_\text{SL} - \gamma_\text{SV})$ is the *work of immersion*, because it quantifies the energy change when a unit area of solid/vapor interface is replaced by solid/liquid interface,

$$W_\text{I} = \gamma_\text{SL} - \gamma_\text{SV} \qquad (3.17)$$

In this case γ_SV may still be determined from Equation (3.16), but the value of h will be negative.

Adamson and Gast [6] have discussed corrections to the capillary-rise technique when the meniscus is not spherical as well as the experimental issues associated with the measurements.

3.2.4 Small droplets

The derivation of Young's equation neglects the *line tension* τ associated with the triple line where the three phases intersect. This introduces no error for larger droplets but for the case of small droplets, namely those less than one micrometer in diameter, the line tension must be considered. In this case Equation (3.5) is written

$$g_{LV}^S \left(\cos\theta_Y \frac{dA_{SL}}{dr} \right) + g_{SV}^S \left(-\frac{dA_{SL}}{dr} \right) + g_{SL}^S \frac{dA_{SL}}{dr} + \tau \frac{dC}{dr} = 0 \tag{3.18}$$

where C is the circumference of the triple line and

$$\frac{dC}{dr} = \frac{1}{r}\frac{dA_{SL}}{dr} \tag{3.19}$$

Therefore, the wetting angle will be expressed as

$$\cos\theta_Y = \frac{\gamma_{SV} - \gamma_{SL}}{\gamma_{LV}} - \frac{\tau}{r\gamma_{LV}} \tag{3.20}$$

3.2.5 Non-ideal surfaces

In reality, solid surfaces are never truly, flat and homogeneous, as assumed in the derivation of Young's equation. Deviation from ideality due to surface roughness and chemical inhomogeneity will now be considered.

3.2.5.1 Rough surfaces

Surface roughness can change the wetting phenomena in two ways. First, roughness will increase the area of contact between solid and liquid. If the true surface area is denoted A_R, and the surface area for a perfectly flat surface is A_I, then the change in free energy as a result of the increase of triple-line length, Equation (3.5), may be expressed as

$$g_{LV}^S \left(\cos\theta \frac{dA_{SL}}{dr} \right) + \frac{A_R}{A_I} \left[g_{SV}^S \left(-\frac{dA_{SL}}{dr} \right) + g_{SL}^S \frac{dA_{SL}}{dr} \right] = 0 \quad (3.21)$$

which yields, in place of Equation (3.7), for the *observed contact angle*

$$\cos\theta_{obs} = \frac{A_R}{A_I} \left[\frac{\gamma_{SV} - \gamma_{SL}}{\gamma_{LV}} \right] = \frac{A_R}{A_I} \cos\theta_Y \quad (3.22)$$

which is known as the Wenzel equation [7]. The Wenzel equation indicates that roughness will decrease wetting for systems with $\theta_Y > 90°$ and increase wetting for those with $\theta_Y < 90°$. For example, for a system with $\theta_Y = 60°$ on a smooth surface, if the surface is roughened to $A_R/A_I = 2$, Equation (3.22) yields

$$\cos\theta_{obs} = 2\cos(60°) = 1.0 \quad (3.23)$$

Thus the observed contact angle will be

$$\theta_{obs} = \cos^{-1}(1.0) = 0° \quad (3.24)$$

and the liquid will perfectly wet the roughened surface. Conversely, for a system with $\theta_Y = 120°$ on a smooth surface, the same roughness will result in $\theta_{obs} = 180°$ and the liquid will be completely non-wetting. This is illustrated in Figure 3.6, where θ_{obs} is plotted versus A_R/A_I for different values of θ_Y.

Rough surfaces can also pin the movement of the triple line, which prevents the droplet from moving to its equilibrium configuration. Such a surface will keep a liquid that is advancing across a surface at an artificially high contact angle, and a liquid that is receding at an artificially low contact angle. How this can occur is illustrated schematically in Figure 3.7 [8]. Surface B is inclined at an angle α with respect to surface

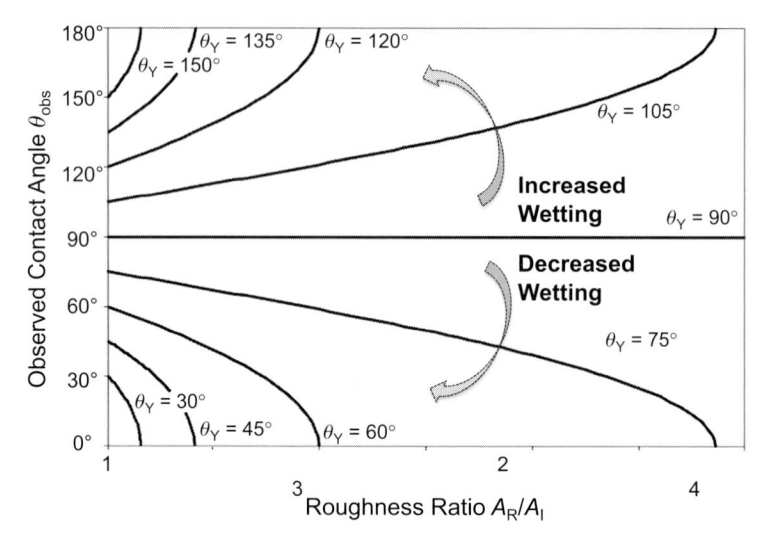

82 EQUILIBRIUM AT INTERSECTIONS OF SURFACES: WETTING

Figure 3.6 The equilibrium contact angle as a function of the surface roughness for different values of θ_Y.

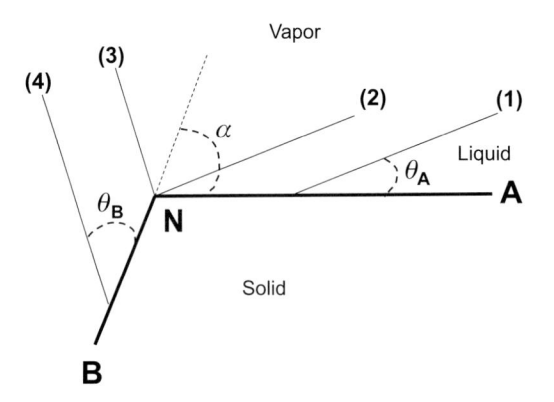

Figure 3.7 The effect of a sharp edge on contact angles when a liquid advances or recedes on a solid surface. From [8], reprinted with permission of John Wiley & Sons.

A. If the liquid resides completely on surface A the contact angle will be θ_Y. However, if additional liquid is added to the droplet, the triple-point junction will move to the left until it reaches the point N, where the contact angle will still be θ_Y and the left-most portion of the liquid/vapor interface will be at position 2. The junction will be pinned at this point until sufficient liquid has been added to the droplet to rotate the interface to position 3, where the contact angle with surface A will be

$$\theta_{Max} = \theta_Y + a \qquad (3.25)$$

Under this condition θ_Y is established on surface B and the junction can proceed as more liquid is added.

Conversely, if the junction lies on surface B, e.g. at position 4, and liquid is removed from the droplet, the junction will move up the surface until it reaches position 3. Now the receding junction will be pinned at N until sufficient liquid has been removed to rotate the liquid/vapor interface to position 2. Thus the contact angle on surface B will vary from θ_Y to θ_{Min}, according to

$$\theta_{Min} = \theta_Y + a \qquad (3.26)$$

Thus on a rough surface the hysteresis of the macroscopic contact angle will vary from θ_{Min} (receding liquid) to θ_{Max} (advancing liquid). Experimentally it is observed that the hysteresis range ($\theta_{Max} - \theta_{Min}$) is only a few degrees for moderate surface roughness ($R_a \approx 100$ nm), but for some non-wetting liquids on very rough surfaces ($R_a \approx 1$ μm) the difference can be as much as 20° [2].

3.2.5.1.1 Composite wetting As a surface is made successively rougher the hysteresis range increases but then drops to virtually zero. If a surface is very rough, a liquid for which $\theta_Y > 90°$ will not penetrate into the valleys but will ride on the peaks as shown schematically in Figure 3.8 [8]. This is known as *composite wetting*. Since the liquid contacts the solid only at a small number of points there is minimal resistance to its motion. The mobility of mercury droplets on ceramic surfaces has been attributed to this effect [2].

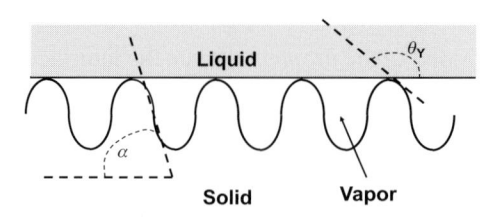

Figure 3.8 The formation of a composite surface for a poorly wetting liquid. From [8], reprinted with permission of John Wiley & Sons.

Eustathopoulos *et al.* [8] show using a model triangular groove geometry that the transition to composite wetting occurs for surfaces such that

$$\alpha > 180° - \theta_Y \qquad (3.27)$$

Composite wetting takes place on naturally occurring surfaces such as plant leaves for which wetting by liquid water is retarded (hydrophobicity) by the natural leaf texture [9].

3.2.5.2 Heterogeneous surfaces

The manner in which a liquid wets a flat, heterogeneous surface is a function of the phase distribution on the multiphase surface. The problem of a liquid wetting a two-phase heterogeneous solid was first addressed by Cassie [10]. Additionally, it is possible for metastable configurations to develop, depending on the solid microstructure and whether the liquid is advancing or receding toward equilibrium, similarly to the effects described for roughness.

A simple example, which can be used to illustrate the wetting behavior of liquids on heterogeneous solids, is a sessile drop on a two-phase surface. Assume the sessile drop in Figure 3.2 covers a surface consisting of phases α and β whose size is small compared with the size of the drop. Let f_A^α and f_A^β be the area fractions of the two phases, respectively. In this case an incremental change in the interfacial area results in a free-energy change of

$$\frac{dG'}{dr} = g_{LV}^S \frac{dA_{LV}}{dr} + \left(f_A^\alpha g_{\alpha V}^S + f_A^{S\beta} g_{\beta V}^S\right) \frac{dA_{SV}}{dr}$$
$$+ \left(f_A^\alpha g_{\alpha L}^S + f_A^\beta g_{\beta L}^S\right) \frac{dA_{SL}}{dr} = 0 \quad (3.28)$$

Use of Equations (3.3) and (3.4) results in

$$g_{LV}^S \left(\cos\theta \frac{dA_{SL}}{dr}\right) + \left(f_A^\alpha g_{\alpha V}^S + f_A^\beta g_{\beta V}^S\right)\left(-\frac{dA_{SL}}{dr}\right)$$
$$+ \left(f_A^\alpha g_{\alpha L}^S + f_A^\beta g_{\beta L}^S\right) \frac{dA_{SL}}{dr} = 0 \quad (3.29)$$

which, upon division through by dA_{SL}/dr, may be rearranged to

$$\cos\theta = f_A^\alpha \frac{g_{\alpha V}^S - g_{\alpha L}^S}{g_{LV}^S} + f_A^\beta \frac{g_{\beta V}^S - g_{\beta L}^S}{g_{LV}^S} \quad (3.30)$$

or

$$\cos\theta_C = f_A^\alpha \cos\theta_\alpha + f_A^\beta \cos\theta_\beta \quad (3.31)$$

This relationship is known as Cassie's law [10].

Cassie's law was derived by minimization of the free energy to determine the equilibrium configuration. In the above derivation, it was assumed that the spatial distribution of α and β phases was uniform. However, if the distribution of phases is not homogeneous, the change in energy that results from the movement of the triple line toward a given location will not necessarily be identical to the energy change that results from the movement of the triple line in the opposite direction. Under these circumstances, a receding liquid can form a contact angle different from that formed by an advancing liquid. This is another example of contact-angle hysteresis.

The concept of contact-angle hysteresis and metastable configurations can be illustrated, following the example of Eustathopoulos et al. [8] by

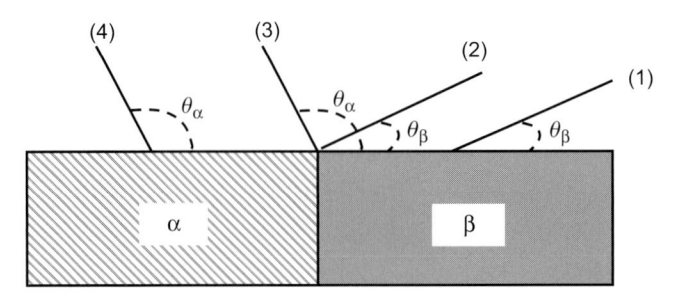

Figure 3.9 Successsive configurations of a liquid spreading on a heterogeneous surface consisting of an α phase that is poorly wet and a β phase that is well wet. From [8], reprinted with permission of John Wiley & Sons.

considering Figure 3.9 for a two-phase system in which the β phase is well wet by the liquid while the α phase is poorly wet. If the liquid resides completely on the β phase the contact angle will be θ_β. If additional liquid is added to the droplet, the triple-point junction will move to the left until it reaches the α/β interface, where the contact angle will still be θ_β and the leftmost portion of the liquid/vapor interface will be at position 2. The junction will be pinned at this point until sufficient liquid has been added to the droplet to rotate the interface to position 3, where the contact angle will be θ_α. Under this condition θ_Y is established on the α phase and the junction can proceed as more liquid is added.

Conversely, if the junction lies on the α phase, e.g. at position 4, and liquid is removed from the droplet the junction will move up the surface until it reaches position 3. Now the receding junction will be pinned until sufficient liquid has been removed to rotate the liquid/vapor interface to position 2. Figure 3.10 shows the variation in contact angle for an advancing and receding liquid as a function of the overall area fraction of α and β for a hypothetical system.

Horsthemke and Schröder [11] have calculated the advancing and receding contact angles for a liquid on a two-phase solid consisting of an

3.3 REACTIVE WETTING 87

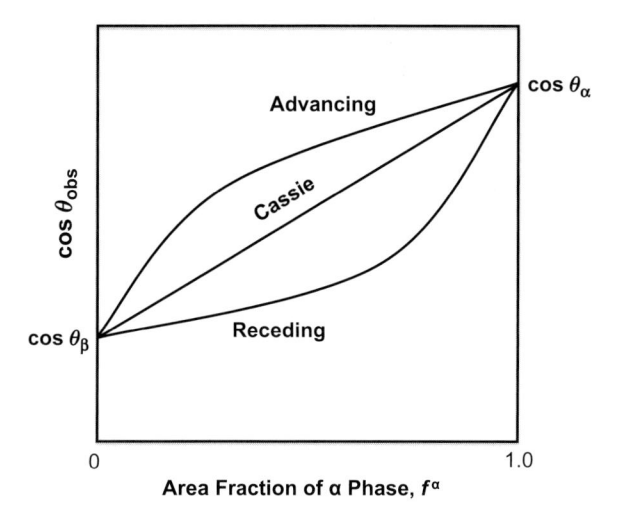

Figure 3.10 The contact angle of an advancing, equilibrium and receding liquid on a two-phase solid consisting of α and β, where $\theta_\alpha \neq \theta_\beta$.

array of hexagons of α and β phases. The results of these calculations are

$$\cos\theta_{obs}^{Adv} = f_{max}^\alpha \cos\theta_\alpha + (1 - f_{max}^\alpha)\cos\theta_\beta \quad (3.32)$$

$$\cos\theta_{obs}^{Rec} = f_{min}^\alpha \cos\theta_\alpha + (1 - f_{min}^\alpha)\cos\theta_\beta \quad (3.33)$$

where f_{min}^α is the local minimum, and f_{max}^α is the maximum area fraction a line along a given wetting direction could cut. Eustathopoulos et al. [2] have shown that Equations (3.32) and (3.33) provide reasonable agreement with the limited data which are available for the spreading of liquids on multiphase solids.

3.3 Reactive wetting

In this section the additional features involved in reactive wetting will be introduced. The reader is referred to the excellent monograph by Eustathopoulos et al. [2] for an in-depth discussion of this subject.

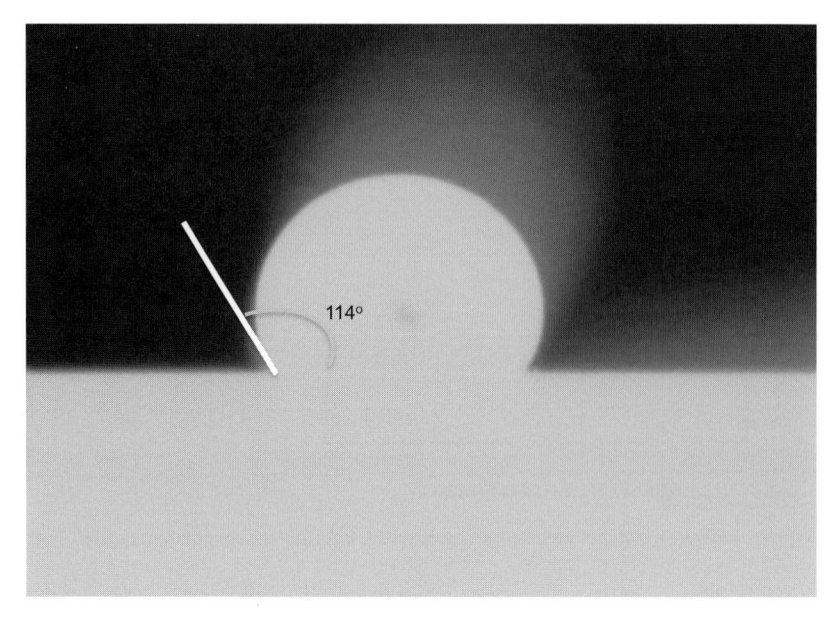

Figure 3.11 A sessile drop of liquid copper on alumina at 1,100 °C. From [12], reprinted with permission of Dr. R. W. Jackson.

The phenomena which occur in reactive wetting are determined by the extent of reaction between the coexisting phases. These phenomena will be illustrated for the case of liquid metals on solid oxide substrates. Jackson [12] studied the wetting of copper on Al_2O_3, yttria-stabilized zirconia (YSZ) and NiO at 1,100 °C. The atmosphere used was a mixture of argon, H_2 and H_2O in which the oxygen partial pressure was 5.65×10^{-7} atm, which is below the dissociation pressure of Cu_2O but above the dissociation pressures of the oxides which were used as substrates (see Figure 1.13).

Figure 3.11 shows an in-situ photograph of a droplet of liquid Cu on alumina, for which the contact angle was $114.5 \pm 0.4°$. The contact angles for the various substrates are listed in Table 3.1. There is limited

Table 3.1 *Contact angles for sessile drops of Cu on oxide substrates [12]*

Liquid	Solid	Contact angle (degrees)
Cu	YSZ	119.7
Cu	Al_2O_3	114.5
Cu	NiO	78.7
Cu–1 wt% Ni	NiO	54.4

wetting of the copper on the YSZ and alumina substrates. The high stability of these oxides results in minimal interaction (e.g. dissolution) with the liquid and the weak bonding results in high values of γ_{SL} and high contact angles. The contact angle on NiO is much lower (78.7°), indicating lowering of γ_{SL} by interaction with the NiO, which is only moderately more stable than Cu_2O (Figure 1.13). The literature contains two explanations for this effect [2]:

(i) dissolution of the solid into the liquid; and
(ii) formation of metal–oxygen clusters in the liquid, which increase the bonding to the solid oxide.

Indeed, dissolution was indicated by a roughening of the Cu/NiO interface. However, as can be seen in Figure 3.12, addition of 1% Ni to the Cu resulted in less dissolution but a still lower contact angle (54.4°). From this observation Jackson [12] concluded the formation of metal–oxygen clusters in the liquid was the reason for the increased wetting.

An additional factor that can affect reactive wetting is the formation of a new phase at the interface between the solid and liquid. If the new phase is wetted by the liquid, the contact angle will decrease. If the new phase is poorly wetted, the contact angle will increase. See Reference [2] for an extensive discussion of these phenomena.

Table 3.2 *Selected values of interfacial energies for solid/liquid and solid/solid interfaces*

Interface	T (°C)	γ (mJ/m^2)	Reference
Al (solid/liquid)	600	93	[13]
Cu (solid/liquid)	1,083	235	[2]
Au (solid/liquid)	1,063	132	[13]
Fe (solid/liquid)	1,535	204	[13]
Ni (solid/liquid)	1,455	255	[12]
Cu (liquid)/W (solid)	1,500	1,120	[2]
Au (solid)/Al$_2$O$_3$ (solid)	1,000	1,725	[13]
Ag (solid)/Al$_2$O$_3$ (solid)	700	1,630	[13]
Cu (solid)/Al$_2$O$_3$ (solid)	850	1,925	[13]
Ni (solid)/Al$_2$O$_3$ (solid)	1,000	2,140	[13]
Pt (solid)/Al$_2$O$_3$ (solid)	1,400	1,050	[13]

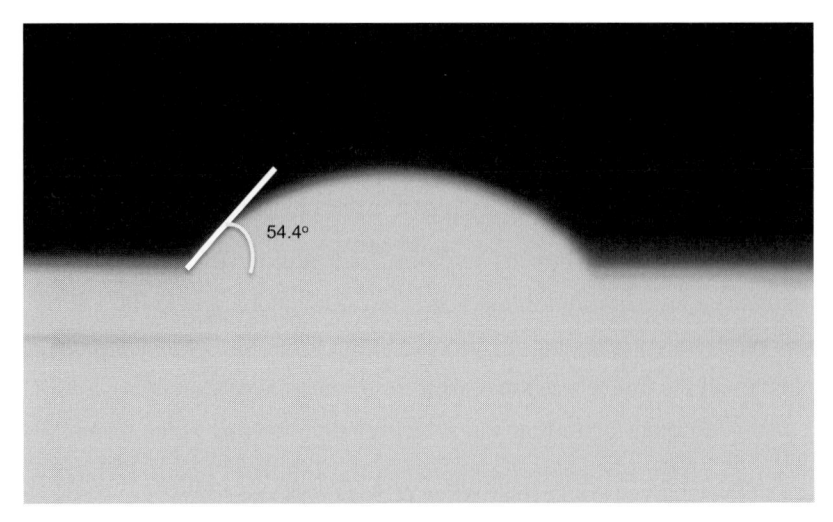

Figure 3.12 A sessile drop of liquid Cu–1% Ni on NiO at 1,100 °C. From [12], reprinted with permission of Dr. R. W. Jackson.

3.4 Selected values of interfacial energies

Selected values of the surface energies of liquids and solids were presented in Chapter 2. Table 3.2 presents selected values of interfacial energies involving solids and liquids.

3.5 Summary

In this chapter the phenomenon of wetting has been described. The fundamental equations have been developed for the case of non-reactive wetting on ideal perfectly smooth surfaces. The complications introduced by non-ideal surfaces, including rough surfaces and heterogeneous surfaces have been described. Finally, the effects of reactive wetting have been introduced.

3.6 References

[1] T. Young, An essay on the cohesion of fluids, *Phil. Trans. Roy. Soc. Lond.*, **94** (1805), 65–87.
[2] N. Eustathopoulos, M. G. Nicholas and B. Drevet, *Wettability at High Temperatures* (Kidlington: Elsevier Science, 1999).
[3] Y. Liu and R. M. German, Contact angle and solid–liquid–vapor Equilibrium, *Acta Mater.*, **44** (1996), 1657–1663.
[4] J. P. Garandet, B. Drevet and N. Eustathopoulos, On the validity of Young's equation in the presence of gravitational and other force fields, *Scripta Mater.*, **38** (1998), 1391–1397.
[5] A. Dupré, *Théorie mécanique de la chaleur* (Paris: Gauther-Villars, 1869), Chapter IX.
[6] A. W. Adamson and A. P. Gast, *Physical Chemistry of Surfaces*, 6th edn. (New York: John Wiley & Sons, 1997), Chapter 1.
[7] R. N. Wenzel, Resistance of solid surfaces to wetting by water, *Ind. Eng. Chem.*, **28** (1936), 988–994.
[8] N. Eustathopoulos, B. Drevet, S. Brandon and A. Virozub, Basic principles of capillarity in relation to crystal growth, in *Crystal*

Growth Processes Based on Capillarity, ed., T. Duffar (Chichester: John Wiley & Sons, 2010), Chapter 1, pp. 1–49.
[9] J. A. Nychka and M. M. Gentleman, Implications of wettability in biological materials science, *JOM*, **62** (2010), 39–48.
[10] A. B. D. Cassie, Contact angles, *Discuss. Faraday Soc.* **3** (1948), 11–16.
[11] A. Horsthemke and J. J. Schröder, The wettability of industrial surfaces: contact angle measurements and thermodynamic analysis, *Chem. Eng. Process.*, **19** (1985), 277–285.
[12] R. W. Jackson, *The Degradation of Refractory Ceramics by Molten Metals*, Ph. D Thesis, University of Pittsburgh, PA, 2010.
[13] L. E. Murr, *Interfacial Phenomena in Metals and Alloys* (Reading, MA: Addison-Wesley Publishing Co., 1975), Chapter 3.

3.7 Study problems

1. Consider the melting of a small piece of iron on an alumina plate at 1,850 °C. Under these conditions the interfacial energy between Fe and Al_2O_3 is 2,000 mJ/m². What will the contact angle be? Take the necessary additional data from Tables 2.1 and 2.2 and assume $d\gamma_{LV}/dT = -0.45$ mJ/(m² K) for iron. Draw a simple sketch to define your angle.

2. Consider the surface energies and surface stresses for the hypothetical Lennard-Jones substance in Table 2.3.
   (a) Calculate the contact angle when a drop of liquid is placed on a {111}-oriented solid at the triple point.
   (b) Calculate the stresses in the solid relative to axes parallel and perpendicular to the plane of the solid.
   (c) Repeat the calculations for a {100}-oriented surface.

3. A pure liquid with a density of 2,000.0 kg/m³ is held in a furnace in a laboratory which is near sea level. A capillary tube (of inner diameter 1.0 mm) is introduced into the metallic bath. The liquid completely wets the inner walls of the tube and the bottom of the

meniscus (approximately spherical) is measured to be at a height of 20.0 cm above the surface of the liquid alloy. Calculate the surface energy of the liquid. Use $g = 9.81$ m/s².

4. When a water droplet is placed on a soda-lime-silica glass it completely wets the glass. When mercury is placed on the same glass it is relatively non-wetting, with $\theta_Y = 150°$. Consider the case where glass capillary tubes with an inner diameter of 0.8 mm are inserted into baths of water and Hg.
   (a) Sketch the resultant morphology in both tubes.
   (b) Calculate the height of the two liquids in the tubes relative to the surface of the bulk liquids.
   Use $\rho_{H_2O} = 998.2$ kg/m³, $\gamma_{H_2O} = 72$ mJ/m², $\rho_{Hg} = 13{,}226.2$ kg/m³ and $\gamma_{Hg} = 485$ mJ/m².

5. Consider the contact-angle data for liquid Cu on various oxides in Table 3.1. Assume a solid block is prepared by plasma spraying alternating layers of NiO and YSZ, with each layer being 50 μm thick. The block is then sectioned normal to the layers to reveal a "striped" surface. Assume a drop of Cu is placed on this surface and heated to 1,100 °C in an Ar–H₂–H₂O atmosphere. Sketch the shape of the drop and calculate the contact angle for directions parallel and perpendicular to the stripes. (The Cu droplet is large enough to cover many stripes.)

6. Consider droplets of liquid mercury on an oxide substrate. Assume the following surface energies: $\gamma_{Hg} = 485$ mJ/m², $\gamma_{Oxide} = 1{,}500$ mJ/m² and $\gamma_{Hg/Oxide} = 1{,}800$ mJ/m².
   (a) Calculate the equilibrium contact angle θ_Y for a large droplet of Hg.
   (b) Assuming an energy for the triple line of 1×10^{-5} mJ/m calculate the contact angles for droplets with radii of 50, 100 and 1,000 nm.

4 Surfaces of crystalline solids

In this chapter the special features of surfaces and internal boundaries in single-phase crystalline systems are described. The orientation dependence of the surface functions will be discussed and the use of Wulff plots and stereographic triangles to present the orientation dependence of surface energy will be demonstrated. The various types of internal boundaries (*grain boundaries*, *twin boundaries*, etc.) and their thermodynamic properties will be discussed.

As described in Chapter 2, there are two fundamental quantities that describe surfaces and interfaces. One is the *surface energy* (γ), which is the reversible work involved in creating unit area of new surface at constant temperature, volume and total number of moles (J/m^2). The other is the *surface stress* (σ), which is the work involved in *reversibly* deforming a surface (N/m). We have seen that for pure liquids the two quantities are numerically equal. The phenomena described in this chapter relate mainly to the surface energy.

4.1 Surface energy for crystalline solids

The surface energy is a scalar quantity, which is isotropic for liquids but is a function of crystallographic orientation for surfaces involving crystalline solids, i.e. it is a function of the surface-plane normal. This was shown qualitatively in Figure 2.3. This can be seen quantitatively for the hypothetical two-dimensional crystal depicted in Figure 4.1 following Mullins [1]. In this example "edges" replace "surfaces". If we assume the atoms are square, with dimensions a on a side, and bonded only to

4.1 SURFACE ENERGY FOR CRYSTALLINE SOLIDS

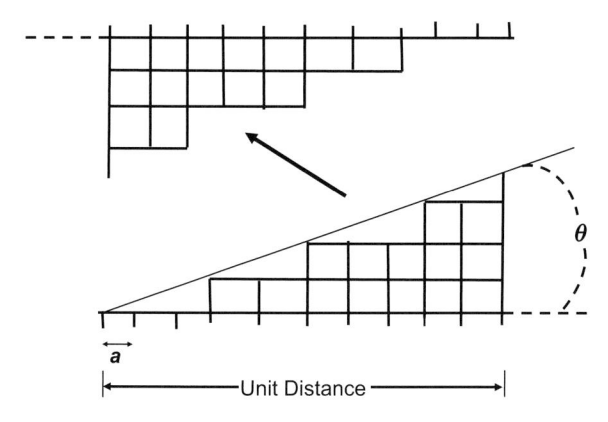

Figure 4.1 A schematic diagram of the formation of a "surface" in a two-dimensional crystal with square atoms with sides of length a. (From [1], reprinted with permission of ASM International.)

nearest neighbors we can express the surface energy (line energy in this case) for different orientations by counting the number of bonds broken in forming the surface, i.e. using the *quasichemical model*, which is described in Section 1.2.2.

The "surface" is created by cutting the crystal along a line at an angle of θ to the horizontal. The number of bonds directed vertically per unit projected length is

$$N_{\text{Vert}} = \frac{1}{a} \tag{4.1}$$

The number of bonds directed horizontally per unit projected length is

$$N_{\text{Horiz}} = \frac{\tan \theta}{a} \tag{4.2}$$

Thus the total number of broken bonds per unit projected length will depend on θ as

$$N(\theta) = \frac{1}{a} + \frac{1}{a} \tan \theta \tag{4.3}$$

and the line energy (analogous to surface energy) can be expressed by multiplying $N(\theta)$ by the bond energy and dividing by the length of the

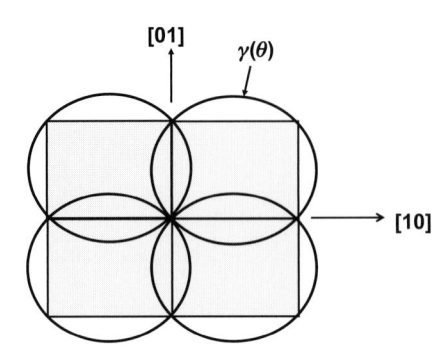

Figure 4.2 A polar plot of the "surface energy" as a function of orientation for the two-dimensional crystal in Figure 4.1. (From [1], reprinted with permission of ASM International.)

surface (analogous to surface area):

$$\gamma(\theta) = \frac{N(\theta)E_{AA}}{L(\theta)} = \frac{(1/a)(1+\tan\theta)E_{AA}}{2/\cos\theta} \quad (4.4)$$

Note that the factor of 2 arises in $L(\theta)$ because two surfaces are being created by this process. We have also

$$\gamma(\theta) = \frac{E_{AA}}{2a}(1+\tan\theta)\cos\theta = \frac{E_{AA}}{2a}(\cos\theta + \sin\theta) \quad (4.5)$$

which may also be written

$$\gamma(\theta) = \frac{E_{AA}}{a\sqrt{2}}\cos\left(\theta - \frac{\pi}{4}\right), \qquad 0 < \theta < \frac{\pi}{4} \quad (4.6)$$

This is the equation of a circle with its center in the first quadrant and passing through the origin of coordinates. By symmetry there is an equivalent circle in each quadrant as shown in Figure 4.2. Such a polar plot, which shows the surface energy as a function of orientation, is known as a Wulff plot [2]. In three dimensions the Wulff plot is generated by rotating a vector through all *hkl* orientations and letting the vector length vary with the magnitude of the surface energy.

The three-dimensional Wulff plot is cumbersome to use. A more convenient approach is to use a stereographic projection. If a crystal is located at the center of a sphere, the normal to each *hkl* plane will intersect the sphere at a particular location or *pole*. The angles

4.1 SURFACE ENERGY FOR CRYSTALLINE SOLIDS

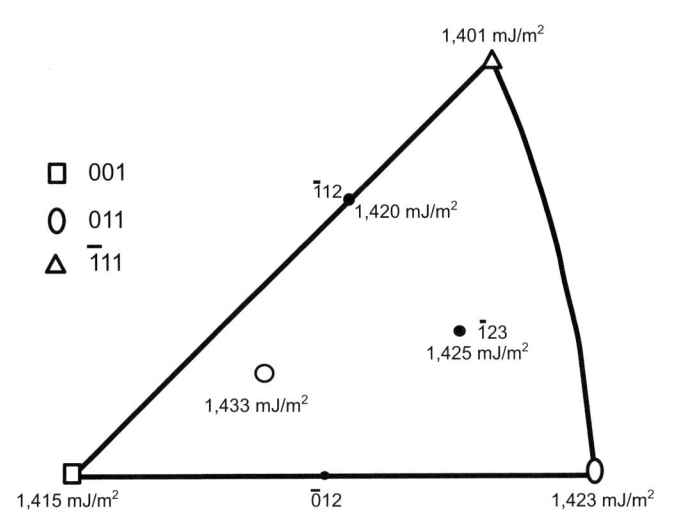

Figure 4.3 A stereographic triangle for a cubic crystal showing selected "poles". Values of the surface free energy for Cu from Reference [4] for selected orientations are superposed on the triangle.

between the normals and hence between the planes can be directly measured on the sphere. The two-dimensional stereographic projection is made by placing a plane tangent to the sphere at one diameter and then projecting the points on the sphere from the other end of the diameter. The projection of the sphere then becomes a circle. Typically the pole for a low-index orientation, e.g. 100, is used as the point of projection. The projection of the various poles allows such information as the angles between planes to be presented. The reader is referred to the text by Cullity and Stock [3] for a clear description of the construction and use of stereographic projections. The symmetry of cubic crystals allows the orientation relations to be represented by a small part of the stereographic projection called a stereographic triangle. Figure 4.3 is such a triangle, from a (100) projection, whose corners are the 001, 011 and $\bar{1}11$ poles. Several other poles are also marked on the triangle.

Table 4.1 *Values of γ_{SV} for Cu for several temperatures and orientations* [4]

Temperature (°C)	γ_{100} (mJ/m²)	γ_{111} (mJ/m²)	γ_{110} (mJ/m²)
827	1,448	1,434	1,477
927	1,415	1,401	1,423
1,027	1,382	1,376	1,394

Data for the surface energies of copper for different orientations were measured at 1,100, 1,200 and 1,300 K by McLean [4]. This was done using a *grain-boundary-grooving technique*, which will be described later. This technique gives only the ratio of surface energies for different orientations. Using $\gamma_{100} = 1{,}415$ mJ/m² at 1,200 K the ratios could be converted to absolute values for other surface orientations. Also, using $d\gamma_{100}/dT$ from an estimated value of the surface entropy (s^S) for (100) surfaces of 0.33 mJ/(m² K), the normalized values at other temperatures were quantified. Values for three orientations are presented in Table 4.1. The values for 1,200 K are superposed on the stereographic triangle in Figure 4.3 along with several other values extracted from contour plots in Reference [4]. There are small cusps at the (100) and (111) poles, and the maximum value of γ occurs fairly near the center of the triangle (the location is indicated by an open circle).

The surface entropies for various orientations were also calculated in [4], from $d(\gamma_{hkl}/\gamma_{100})/dT$. This function had minimum values at the (100) and (111) poles and maximum values near the (210) pole.

4.1.1 Equilibrium crystal shape

The relative magnitudes of the surface free energies for different crystallographic planes will determine the shape of a crystal if it has time to come to equilibrium. In Figure 4.2 the equilibrium shape of the

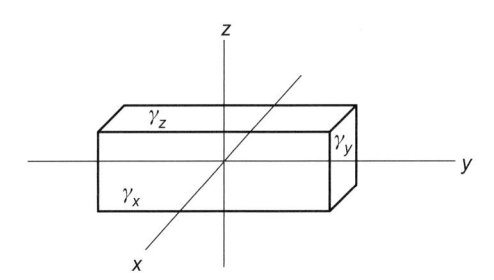

Figure 4.4 A schematic diagram of a rectangular crystal with different values of the surface energy for surfaces normal to the three coordinate directions (x, y and z).

two-dimensional crystal is a square shown superimposed on the Wulff plot. Even though the square has a larger surface area (line length) than a circle, it has lower total surface energy because its surface is constructed entirely of low-γ planes (edges).

This concept may be extended to three dimensions. Minimum free energy requires that $F^S = \int \gamma(\theta) dA$ be a minimum. For a liquid $F^S = \gamma \int dA = \gamma A$ and the equilibrium shape is a sphere. This will, in general, not be the case for a crystal. Minimization of F^S for a general crystal shape is a relatively complex calculation. Therefore, in the interest of simplicity, following Mullins [1] a calculation for a crystal with fixed orientation of its surfaces will be performed. Consider the crystal in Figure 4.4, which exists in equilibrium with its vapor. It will be assumed that the geometry of the crystal remains that of a rectangular solid. The surface energies for each face will be identified by the direction of the plane normal, i.e. γ_x, γ_y and γ_z. Each face will be at a distance l_i ($i = x$, y and z) from the center of coordinates. For a rectangular crystal the dimensions are $2l_x$, $2l_y$ and $2l_z$. The equilibrium shape (dimensions) will be that which minimizes F^S for a fixed volume given by

$$V' = (2l_x \cdot 2l_y \cdot 2l_z) \tag{4.7}$$

The total free energy associated with the surfaces is given by

$$F^S = 2\gamma_x(2l_y \cdot 2l_z) + 2\gamma_y(2l_x \cdot 2l_z) + 2\gamma_z(2l_x \cdot 2l_y) \tag{4.8}$$

The minimization of F^S requires

$$dF^S_{T,V} = 0 = 2\gamma_x\left[2l_y\,d(2l_z) + 2l_z\,d(2l_y)\right] + 2\gamma_y\left[2l_x\,d(2l_z) + 2l_z\,d(2l_x)\right]$$
$$+ 2\gamma_z\left[2l_x\,d(2l_y) + 2l_y\,d(2l_x)\right] \qquad (4.9)$$

subject to the constraint of constant volume. This problem of a "constrained minimum" may be solved by the *method of undetermined multipliers*:

$$-\lambda\,dV' = 0$$
$$= -\lambda\left[(2l_y \cdot 2l_z)d(2l_x) + (2l_x \cdot 2l_z)d(2l_y) + (2l_x \cdot 2l_y)d(2l_z)\right] \qquad (4.10)$$

Here λ is an undetermined multiplier to account for one of the crystal dimensions being a dependent variable,

$$dF^S_{T,V} - \lambda\,dV' = 0 \qquad (4.11)$$

Substitution of Equations (4.9) and (4.10) into Equation (4.11) and rearrangement to collect the coefficients of dl_x, dl_y and dl_z yields

$$\left[2\gamma_y \cdot 2l_z + 2\gamma_z \cdot 2l_y - \lambda(2l_y \cdot 2l_z)\right]d(2l_x)$$
$$+ \left[2\gamma_x \cdot 2l_z + 2\gamma_z \cdot 2l_x - \lambda(2l_x \cdot 2l_z)\right]d(2l_y)$$
$$+ \left[2\gamma_x \cdot 2l_y + 2\gamma_y \cdot 2l_x - \lambda(2l_x \cdot 2l_y)\right]d(2l_z) = 0 \qquad (4.12)$$

If l_z is chosen as the "dependent dimension" and the *undetermined multiplier* λ is arbitrarily chosen to make the coefficient of dl_z equal to zero,

$$2\gamma_x \cdot 2l_y + 2\gamma_y \cdot 2l_x - \lambda(2l_x \cdot 2l_y) = 0 \qquad (4.13)$$

Division of Equation (4.13) by $(2l_x \cdot 2l_y)$ yields

$$\frac{\gamma_x}{l_x} + \frac{\gamma_y}{l_y} - \lambda = 0 \qquad (4.14)$$

Now l_x and l_y are independent variables, so their coefficients must also be zero, which leads to

$$\frac{\gamma_x}{l_x} + \frac{\gamma_y}{l_y} = \frac{\gamma_x}{l_x} + \frac{\gamma_z}{l_z} = \frac{\gamma_y}{l_y} + \frac{\gamma_z}{l_z} = \lambda \quad (4.15)$$

which may be simplified to

$$\frac{\gamma_x}{l_x} = \frac{\gamma_y}{l_y} = \frac{\gamma_z}{l_z} = \frac{\lambda}{2} \quad (4.16)$$

which is known as the Wulff equation. Equation (4.16) indicates that those surfaces which have low values of γ will have small values of l, i.e. they will lie close to the center of the crystal and will, therefore, have significant areas. Conversely, those surfaces with high values of γ will have large values of l, i.e. they will lie far from the center of the crystal and will, therefore, have small areas. Thus, the total surface energy of the crystal will be minimized by having most of the surface consist of low-γ planes.

The above treatment embodies the essence of the Wulff construction but is restricted by a prior choice of which planes would contribute to the surface. A more general approach may be used if the complete Wulff plot is available. This treatment was outlined by Herring [5] as follows.

(i) Vectors are drawn from the center to various reperesentative locations on the Wulff plot.
(ii) Lines are constructed normal to the vectors where they touch the Wulff plot.
(iii) The equilibrium shape of the crystal will be given by the "inner envelope" of these planes, i.e. it will be the shape which minimizes $F^S = \int \gamma(\theta) dA$.

This procedure would yield the square shape for the two-dimensional example of Figure 4.2. The Wulff construction is an elegant concept from the standpoint of understanding crystal shapes but is not of significant practical value since the entire Wulff plot is rarely available.

Figure 4.5 An etched surface of polycrystalline nickel showing the grain structure.

4.2 Internal boundaries

The discussion above concerned single crystals. However, most crystalline materials are *polycrystalline*, i.e. they are comprised of an assembly of many crystals, usually referred to as *grains*, which are bonded together but have some degree of crystallographic misorientation relative to one another. The grain boundaries where these grains meet behave in many ways the same as the surfaces described above. Figure 4.5 shows a nickel surface that has been chemically etched to reveal several types of boundaries. The curved boundaries surrounding the rather equiaxed grains are *high-angle grain boundaries* and the very straight boundaries are *twin boundaries*. These will be described below.

4.2.1 Types of grain boundaries

The following is a description of grain boundaries that is based primarily on the degree of misorientation between the adjacent grains.

Small-angle boundaries

Small-angle boundaries are ones across which the difference in orientation is small. They are also called *subgrain boundaries*. There are two general classes of small-angle boundaries: *tilt boundaries*, which comprise an array of edge dislocations, and *twist boundaries*, which comprise two or more arrays of screw dislocations. (The reader who is unfamiliar with the line defects known as *dislocations* will find a clear and readable description in Reference [6].)

Tilt boundaries The formation of a *symmetric tilt boundary* by joining two single crystals together is illustrated in Figure 4.6. Figure 4.6(a) shows two separate crystals of thickness D, each of which has a surface that does not lie along a crystallographic plane of atoms and is inclined at an angle of $\theta/2$ with respect to the vertical atomic planes. The vertical planes will produce edge dislocations when the crystals are joined and, therefore, are labeled with the typical symbol for an edge dislocation ($\perp$). Figure 4.6(b) shows the crystals after they have been rotated and joined. The atomic planes in the two crystals are now misoriented relative to each other by an angle θ. Finally, Figure 4.6(c) shows the boundary after the atoms near the boundary have rearranged themselves to consist of a vertical array of edge dislocations. The vertical spacing between the dislocations is

$$d = \frac{\vec{b}}{\sin \theta} \cong \frac{\vec{b}}{\theta} \tag{4.17}$$

where $\vec{b}$ is the Burgers vector of the dislocations. The energy of the tilt boundary is essentially the sum of the energies of the dislocations in unit area of the boundary. The strain energy associated with each edge dislocation is [6]

$$E_\perp = \frac{\mu b^2}{4\pi(1-\nu)} \log\left(\frac{d}{5\vec{b}}\right) \tag{4.18}$$

(a)

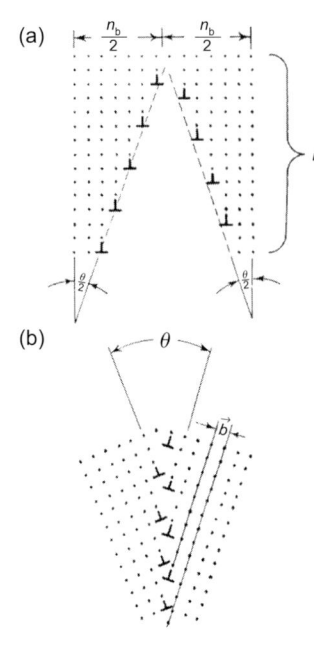

(b)

(c)

Figure 4.6 The formation of a symmetric tilt boundary. (This figure appeared originally in J. Weertman and J. R. Weertman, *Elementary Dislocation Theory* (Oxford: Oxford University Press, 1992) and is reproduced by permission of Oxford University Press.)

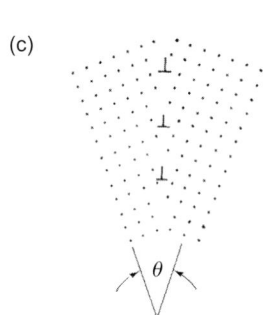

where μ is the shear modulus and ν is Poisson's ratio. The number of dislocations per unit area of boundary is $1/d$, so an approximate expression for the tilt-boundary energy as a function of θ is

$$\gamma_{\text{gb}} \approx \theta \left[\frac{E_c}{\vec{b}} - \frac{\mu b^2}{4\pi(1-\nu)} \log(5\theta) \right] \quad (4.19)$$

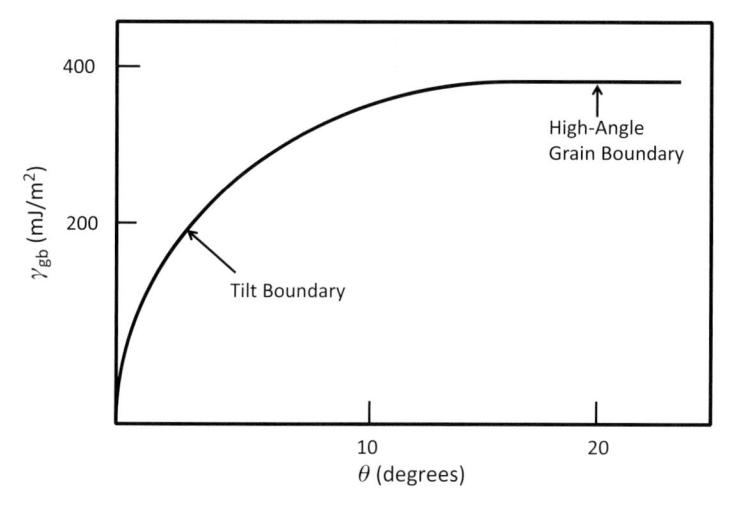

Figure 4.7 A plot showing the variation of grain-boundary energy with tilt angle for symmetric tilt boundaries in a hypothetical metal similar to Ag or Cu.

where E_c is the core energy of the dislocation. This expression is valid only at small misorientation angles ($\theta \approx 10°$). The values of γ_{gb} for a hypothetical metal similar to Ag or Cu are plotted in Figure 4.7. The energy varies significantly at low values of θ, as indicated by Equation (4.19), and then becomes nearly independent of θ at higher angles, for which there are no longer individual dislocations. These boundaries are high-angle grain boundaries.

The geometry of the non-symmetric tilt boundaries which form when the two crystals are not mirror images of each other is more complex and must be described in terms of two arrays of edge dislocations [6].

High-resolution transmission electron microscopy (HRTEM) images of the atomic positions in several tilt grain boundaries are provided in Reference [7].

Twist boundaries A schematic diagram of the formation of a twist boundary is presented in Figure 4.8. A cylindrical single crystal is cut

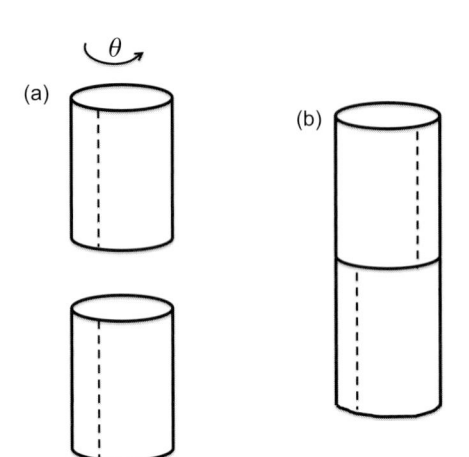

Figure 4.8 A schematic diagram showing the formation of a twist boundary. (This figure appeared originally in J. Weertman and J. R. Weertman, *Elementary Dislocation Theory* (Oxford: Oxford University Press, 1992) and is reproduced by permission of Oxford University Press.)

in two (Figure 4.8(a)). The upper half of the crystal is rotated through a small angle θ and the two halves of the crystal are joined along the plane of the cut (Figure 4.8(b)). If the boundary where the two lattices now meet is a crystallographic plane, the displacements of the atoms are symmetrical with respect to the boundary. This boundary may be considered to comprise two arrays of screw dislocations as illustrated schematically in Figure 4.9. Figure 4.9(a) shows the introduction of one set of vertical screw dislocations onto a plane in a square crystal, which has resulted in a displacement of the two portions of the crystal by an angle θ. Figure 4.9(b) shows the displacement produced by a single set of horizontal screw dislocations. Finally, Figure 4.9(c) shows the effect of combining the two arrays, which produces the rotation necessary for the twist boundary. If each set of screw dislocations contains n dislocations the angle θ is given by

$$\theta = n\vec{b} \qquad (4.20)$$

and the energy of symmetrical twist boundaries varies with θ in a manner similar to that described for tilt boundaries. This may be seen for the case of twist boundaries in silicon (diamond cubic structure) formed

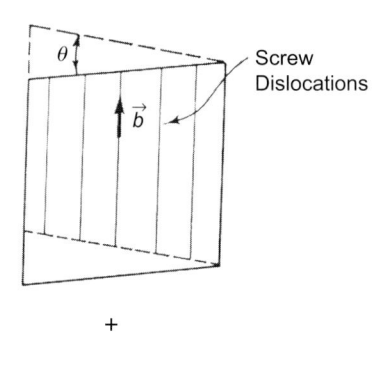

Figure 4.9 A schematic diagram of the networks of screw dislocations that make up a symmetric twist boundary. (This figure appeared originally in J. Weertman and J. R. Weertman, *Elementary Dislocation Theory* (Oxford: Oxford University Press, 1992) and is reproduced by permission of Oxford University Press.)

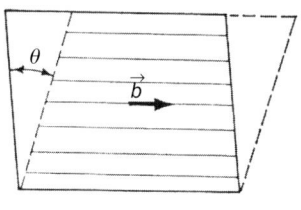

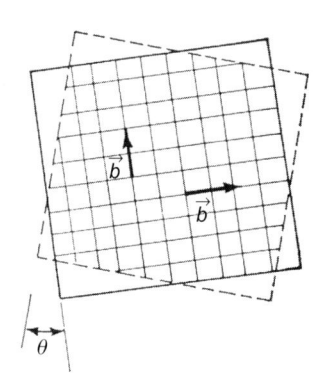

by rotation about a [001] axis in Figure 4.10. (This plot was drawn by combining experimental values of $\gamma_{gb}/(2\gamma_{SL})$ from grain-boundary grooves in silicon in contact with liquid tin in Reference [8] with an estimated value of $\gamma_{SL} = 385$ mJ/m² also from [8].) The value of γ_{gb} increases steeply with θ up to $\theta \approx 10°$, after which it becomes essentially

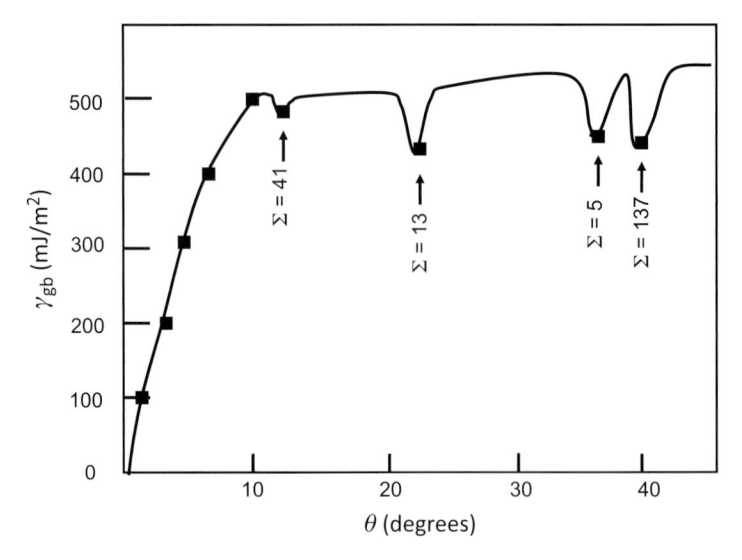

Figure 4.10 A plot of the grain-boundary energy for [001] twist boundaries in silicon versus the twist angle at 1,200 °C.

independent of θ except for special orientations, which will be discussed in the next section.

The geometry of a non-symmetric twist boundary is more complicated, requiring three sets of screw dislocations for its description [6].

High-angle grain boundaries

For misorientations exceeding about 10° it is no longer possible to describe the grain-boundary structure using simple dislocation models. Also, the strain fields of the dislocations start to interact and the dependence of the grain-boundary energy on the misorientation becomes much weaker, as can be seen in Figure 4.10. These *high-angle grain boundaries* are characterized by a high degree of atomic disorder. However, there are some special orientations for which the atomic matching is relatively good.

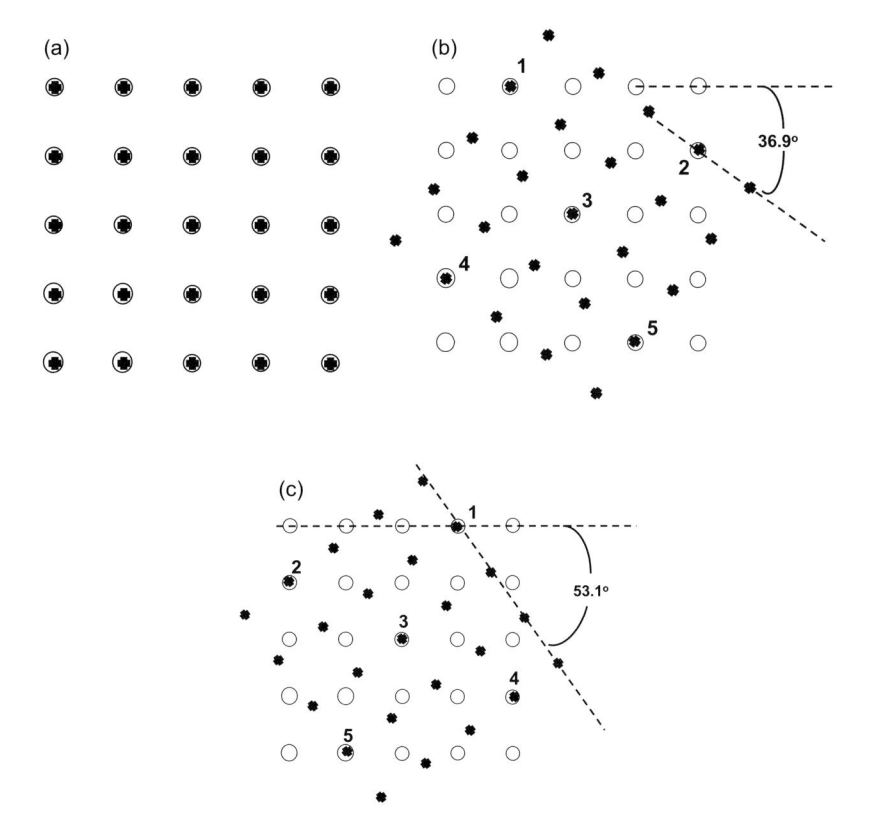

Figure 4.11 A simple-cubic lattice viewed along a [001] direction showing the atom positions in the perfect crystal and how they change with the formation of [001] twist boundaries.

Consideration of the atomic positions in twist boundaries is a convenient way to introduce a methodology for describing such special orientations for any type of grain (or interphase) boundary. Figure 4.11(a) shows 25 atom positions in a simple-cubic lattice viewed along a [001] direction. The circles represent the atoms of the (001) plane below that on which the boundary will be formed, whereas the crosses represent the atoms in the (001) plane above. The twist boundary is formed by rotating the top half of the crystal about the [001] direction, as shown in

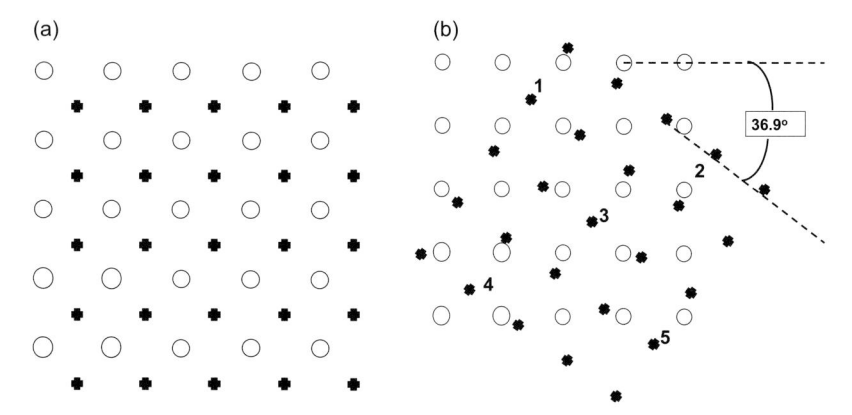

Figure 4.12 A body-centered cubic lattice viewed along a [001] direction showing the atom positions in the perfect crystal and how they change with the formation of [001] twist boundaries.

Figure 4.9. As the rotation begins, the atoms above and below the plane no longer coincide. However, at some specific rotation angles a fraction of the atoms will be brought back into coincidence. Figure 4.11(b) shows the position of the atoms after a rotation of 36.9°. The five numbered atoms (a fifth of the total) now coincide. This behavior can be described in terms of the *coincident-site lattice* (CSL). This is the lattice formed by the points of coincidence. The CSL is described by Σ, which is the inverse of the fraction of the coincident sites. Therefore the CSL for the twist boundary in Figure 4.11(b) is a $\Sigma = 5$ boundary. The complete notation for this boundary is $\Sigma = 5, 36.9°/[001]$.

Figure 4.11(c) shows the same simple-cubic system where the rotation angle has been increased to 53.1°. This rotation results in a $\Sigma = 5$, 53.1°/[001] twist boundary.

Figure 4.12(a) shows a bcc lattice viewed along a [001] direction. Figure 4.12(b) shows the twist boundary produced by rotating the upper portion of the crystal through an angle of 36.9°. The five numbered

4.2 INTERNAL BOUNDARIES

Table 4.2 *Combinations of angle and Σ for cubic crystals for low-Σ boundaries produced by rotations about the ⟨001⟩ and ⟨011⟩ axes*

Axis	Angle (degrees)	Σ	Axis	Angle (degrees)	Σ
⟨001⟩	22.62	13a	⟨011⟩	26.53	19a
	38.07	17a		38.94	9
	36.87	5		50.48	11
	53.13	5		70.53	3
	61.93	17a		86.63	17b
	67.38	13a		93.37	17b
	112.62	13a		109.47	3
	118.07	17a		129.52	11
	126.87	5		141.06	9
	143.13	5		153.47	19a
	151.93	17a			
	157.38	13a			

This table and Figures 4.11 and 4.12 show that different rotation angles can result in boundaries with the same value of Σ. Some boundaries generate only one twin system. Others (e.g. Σ = 13, 17, 19 etc.) can generate two twin systems. These are distinguished by the addition of the letters "a" and "b" to the boundary designation.

atoms above the boundary plane are in the same positions relative to atoms below the plane as before rotation. That is, five lattice sites have been brought into coincidence, so a rotation of 36.9° results in a Σ = 5, 36.9°/[001] twist boundary for the bcc lattice just as it did for the simple-cubic lattice. This is the case for all cubic lattices. Table 4.2 lists the CSL which occur for cubic lattices rotated about the ⟨001⟩ and ⟨011⟩ axes. A more extensive table for cubic crystals is provided in Reference [7].

One might expect the grain-boundary energy to have low values for low values of Σ. However, this is often not the case. This may be seen for [001] twist boundaries in silicon in Figure 4.10 [8]. Four cusps are illustrated for particular values of Σ. (Some of the observed cusps have been omitted from the figure for clarity.) Cusps were observed for $\Sigma = 5$, 13, 17, 25, 37, 41, 61, 85, 89 and 137. However, no cusps were observed at values of θ corresponding to $\Sigma = 29$, 53 and 65. Furthermore the cusp for $\Sigma = 137$ was of the same magnitude as that at $\Sigma = 5$.

Twin boundaries A special type of high-angle boundary is a *twin boundary*. A twin boundary divides two parts of a crystal that are related to each other by a symmetry operation, e.g. the lattice of one part could be the mirror image of the other part reflected across a particular crystallographic plane called the *twinning plane*. The plane of the twin boundary is also called the *composition plane*. Twinning can occur during growth of a crystal, deformation of the crystal or recrystallization following deformation and has been observed both in metals and in inorganic compounds [9].

If the twinning plane and the twin boundary coincide the atoms at the boundary fit perfectly into both lattices. This is illustrated in Figure 4.13(a) [10]. This corresponds to a *coherent twin boundary*. When the twin boundary and twinning plane are not parallel, an *incoherent twin boundary* results and the atomic fit is disrupted; see Figure 4.13(b). The energy of the coherent boundary tends to be quite low because of the good atomic matching, and the energy of the incoherent boundary increases as the angle between the twinning plane and the twin boundary increases, as shown schematically in Figure 4.13(c). Table 4.3 presents approximate values for the surface and several boundary energies for copper.

Figure 4.14 shows an annealing twin in pure nickel. It is to be noted that the intersection of the twin with the grain boundaries results in only a small deflection of the grain boundaries, which is consistent with the low energy of the twin boundary. The intersection of twins with grain boundaries has been analyzed extensively by Murr [12].

Table 4.3 *A comparison of values for several types of boundary in pure Cu*

Type of boundary	γ (mJ/m^2)	Reference
Solid/vapor (average)	1,400	[4]
High-angle grain boundary	450	[11]
Incoherent twin boundary	300	[11]
Coherent twin boundary	20	[10]

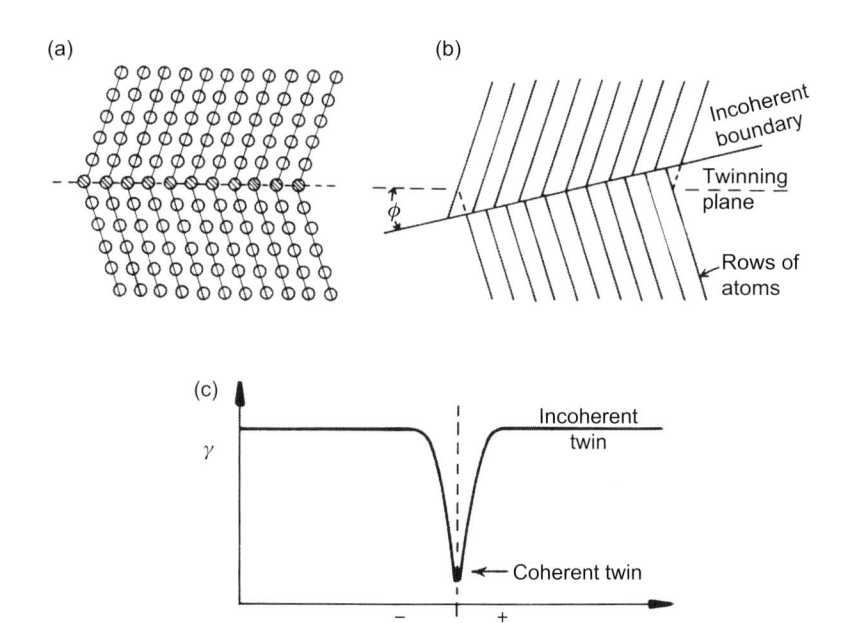

Figure 4.13 A schematic diagram of the formation of twin boundaries. (This figure appeared originally in D. A. Porter and K. E. Easterling, *Phase Transformations in Metals and Alloys* (London: Chapman & Hall, 1992) and is reproduced by permission of Springer Science + Business Media.)

4.2.2 Intersections of grain boundaries with free surfaces

When a grain boundary intersects a free surface it creates a triple point where two solid phases (grains) and the vapor phase meet. (This is analogous to that in Figure 3.1, where a solid phase, a liquid phase

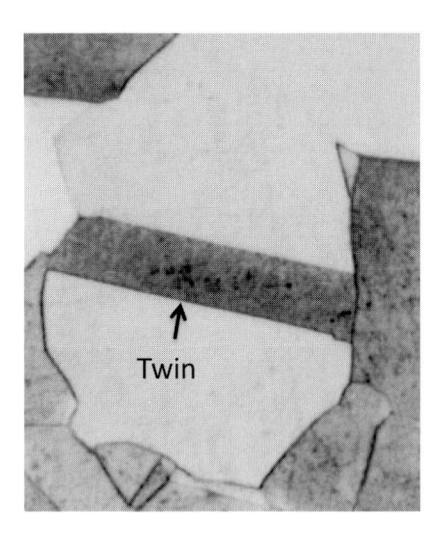

Figure 4.14 A annealing twin in pure Ni.

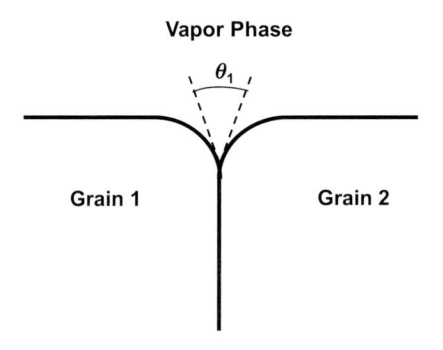

Figure 4.15 A schematic diagram showing the process of *grain-boundary grooving*.

and a vapor phase meet.) The equilibrium morphology is in the form of a groove, as shown in Figure 4.15. The equilibrium dihedral angle θ_1 is determined by a similar type of free-energy minimization to that which was used to calculate the wetting angle θ_Y in Equation (3.7). Consider unit distance into the plane of the page for the triple-point line in Figure 4.15. If this point is reversibly moved a small distance dl, to decrease the length of the grain boundary, the change in grain-boundary area will be dl and the change in the areas of the two solid/vapor interfaces

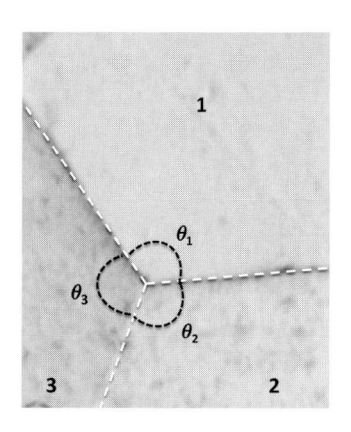

Figure 4.16 A grain-boundary triple point in nickel.

will be $dl \cos(\theta_1/2)$. The total change in free energy may be set equal to zero as follows:

$$2\gamma_{SV}\, dl \cos\left(\frac{\theta_1}{2}\right) - \gamma_{gb}\, dl = 0 \qquad (4.21)$$

Division of Equation (4.21) by dl yields the following expression from which the equilibrium dihedral angle may be calculated:

$$\cos\left(\frac{\theta_1}{2}\right) = \frac{\gamma_{gb}}{2\gamma_{SV}} \qquad (4.22)$$

The values of γ_{SV} for copper for different surface orientations (Table 4.1) were determined from measurements of grain-boundary groove angles in Reference [4].

4.2.3 Intersections of grain boundaries

Figure 4.16 shows the intersection of three grains in annealed nickel with the grain boundaries traced by dashed white lines. The equilibrium angles of intersection of the grain boundaries may be calculated by a similar free-energy minimization to that used in deriving Equation (4.22). If the grain boundaries do not correspond to special boundaries the equilibrium angles will be given by

$$\frac{\gamma_{23}}{\sin\theta_1} = \frac{\gamma_{13}}{\sin\theta_2} = \frac{\gamma_{12}}{\sin\theta_3} \qquad (4.23)$$

If the three boundaries have the same energies, all three angles will be 120°, which is the case for the nickel in Figure 4.16. If any of the boundaries are special high-angle boundaries, with an orientation near to a cusp on the Wulff plot, the calculation is more complex insofar as "torque" terms must be included to account for the tendency of the boundary to rotate into its low-γ orientation [5].

4.3 Faceting

Figure 4.3 indicates that the anisotropy of the surface energy of Cu is small. However, for some solids the anisotropy of γ is significant. For small crystals the equilibrium shape (as in Figure 4.4) will be achieved relatively quickly. However, for large crystals the amount of mass transport required in order to reach the equilibrium shape takes a very long time and a metastable equilibrium can be achieved if macroscopic surfaces break up into *facets* composed of planes of low-γ orientations. This is illustrated for Ag in Figures 4.17 and 4.18. Figure 4.17(a) shows a low-magnification micrograph of a specimen of silver that has been annealed in air at 900 °C. The appearance is somewhat similar to that of the chemically etched Ni specimen in Figure 4.5. However, the Ag specimen has been *thermally etched*. The grains and twins are apparent because of a change in surface topography. This is apparent in the higher-magnification micrograph in Figure 4.17(b) which shows a twin running through a grain that is distinguished by the difference in faceting. Figure 4.18 shows more detail of the faceting of the twin and the neighboring grain.

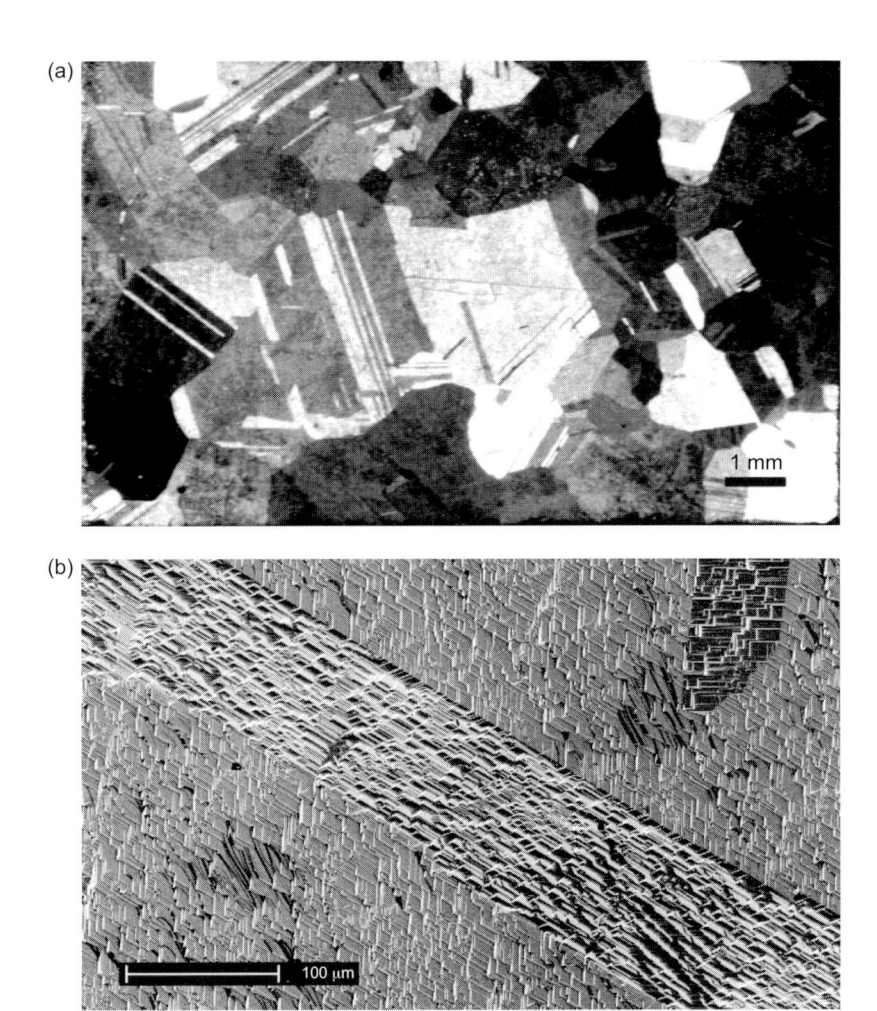

Figure 4.17 Surface micrographs of solid silver showing that a surface with a high-γ orientation will break up into a set of *facets* of lower-γ orientation.

The faceting of the surface of a given grain increases the surface area but actually decreases the contribution of the surface to the free energy of the crystal. This is illustrated in Figure 4.19, which is a schematic diagram showing the process of faceting in two dimensions. Figure 4.19(a)

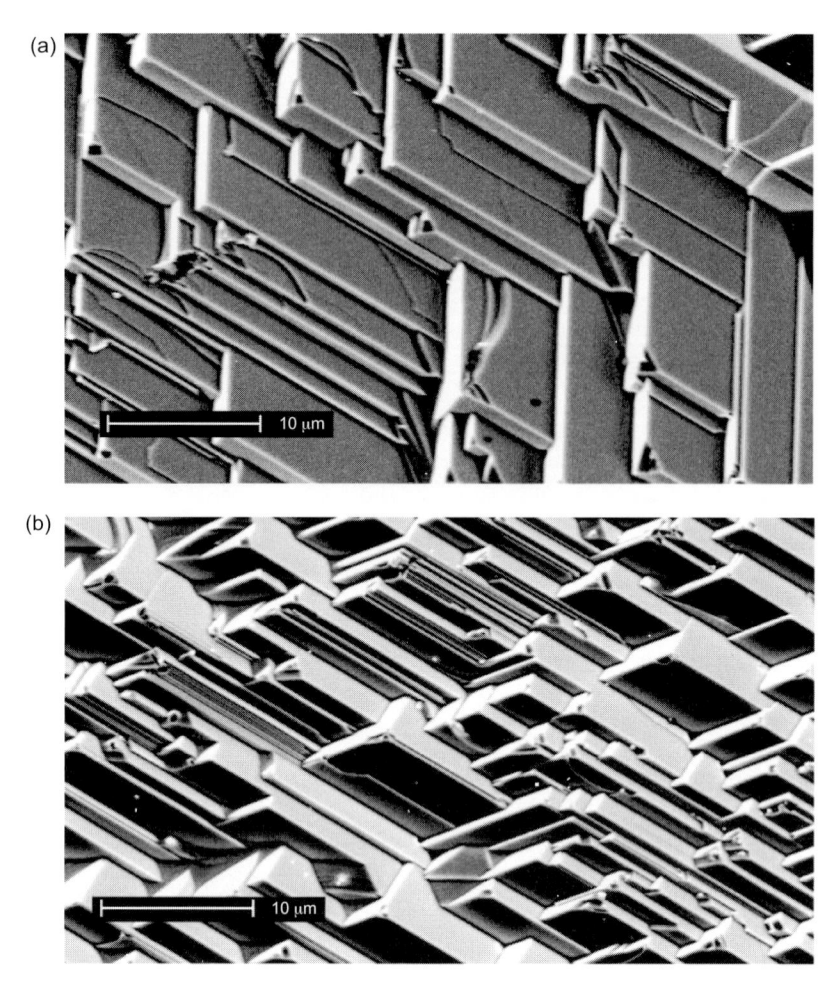

Figure 4.18 Higher-magnification micrographs of the twin (b) and parent grain (a) in Figure 4.17(b).

shows the initial surface, which is formed by a plane with surface energy γ_0 and is at an angle to the low-index planes of the crystal. Figure 4.19(b) shows the crystal after it has rearranged to be constructed of two sets of planes with surface energies γ_1 and γ_2, respectively. Figure 4.19(c) is a schematic Wulff plot for this system, which indicates that planes of

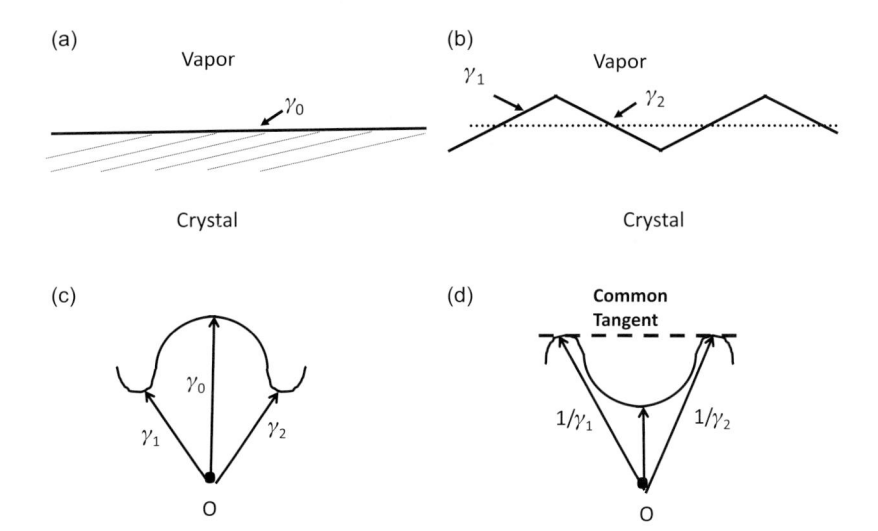

Figure 4.19 A schematic diagram showing the process of *faceting*: (a) the initial surface, with surface energy γ_0 at an angle to the low-index planes of the crystal; (b) the crystal after it has rearranged to two sets of planes with surface energies γ_1 and γ_2; (c) a schematic Wulff plot for this system; and (d) polar plot of $1/\gamma$ for this system.

orientations 1 and 2 have low values of γ. Thus, even though the surface area has increased according to

$$A_1 + A_2 > A_0 \tag{4.24}$$

the surface free energy has decreased:

$$\gamma_1 A_1 + \gamma_2 A_2 < \gamma_0 A_0 \tag{4.25}$$

The relation between the Wulff plot and faceting has been described in detail by Herring [5]. Figure 4.19(d) shows an alternative approach introduced by Meijering [13], who showed that, if the surface energies were plotted on a polar plot as $1/\gamma$, the free-energy balance could be treated using the "common-tangent" approach described for free-energy diagrams in Chapter 1. Thus, one would predict that a surface with orientation 1 could lower its free energy by faceting and the equilibrium orientations of the facets would be given by the common tangent. This approach is particularly helpful in describing faceting in three dimensions such as that shown in Figures 4.17 and 4.18.

4.4 Measurement of surface and grain-boundary energies

4.4.1 The zero-creep technique

The most common method for measuring the surface energy of solids is the zero-creep technique. This technique makes use of the fact that a fine wire, if heated to a temperature high enough to allow significant atomic mobility, will decrease its length by creep in order to decrease its surface area and hence its total surface energy. If a force is applied to the wire, e.g. by hanging a weight from it, the wire will continue to shorten if the force is small and will lengthen if the force is large. For one intermediate value of force the wire will neither shorten nor lengthen. This condition of "zero-creep" can be used to measure the surface energy, γ, from an energy balance. Consider first the simpler case of an amorphous solid, Figure 4.20(a). The energy balance for the creeping wire will be

$$F \, dl = \gamma \, dA_{\text{cyl}} = \gamma (2\pi r \, dl + 2\pi l \, dr) \tag{4.26}$$

Since the deformation is plastic, the volume of the cylinder remains constant, which allows dr and dl to be related,

$$dV = \pi r^2 \, dl + 2\pi r l \, dr = 0 \tag{4.27}$$

which yields

$$dr = -\frac{r \, dl}{2l} \tag{4.28}$$

4.4 MEASUREMENT OF SURFACE AND GRAIN-BOUNDARY ENERGIES

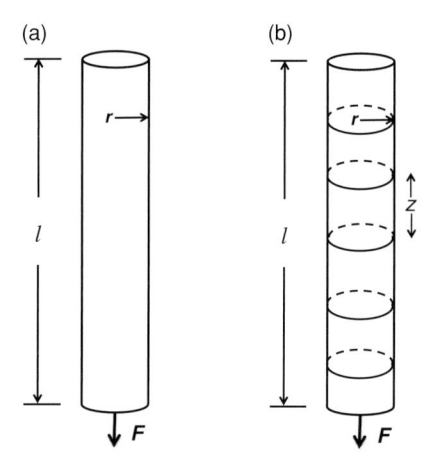

Figure 4.20 A schematic diagram of the zero-creep technique for measuring the surface energy of solids.

Substitution into dA_{cyl} yields

$$dA_{cyl} = \pi r \, dl \tag{4.29}$$

and the energy balance becomes

$$F \, dl = \pi r \gamma \, dl \tag{4.30}$$

and for the force that results in $dl = 0$, i.e. zero creep,

$$\gamma = \frac{F_{zc}}{\pi r} \tag{4.31}$$

For the case of a polycrystalline material the analysis is a little more complex because the grain-boundary energy enters into the balance. Typically when a thin, polycrystalline wire is annealed at high temperature the grain boundaries will rotate until they are normal to the wire axis; see Figure 4.20(b). In this case the energy balance will be

$$F \, dl = \gamma \, dA_{cyl} + \gamma_{gb} \, dA_{gb}^{tot} \tag{4.32}$$

The change in total grain boundary area is

$$dA_{gb}^{tot} = 2n\pi r \, dr \tag{4.33}$$

where n is the number of grain boundaries in the wire, which is related to the spacing of the boundaries by $n = z/l$. Substitution of Equation (4.28) into Equation (4.33) yields

$$dA_{gb}^{tot} = -\frac{n\pi r^2}{l} dl \qquad (4.34)$$

and the energy balance becomes

$$F\, dl = \pi r \gamma\, dl - \frac{n\pi r^2}{l}\gamma_{gb}\, dl \qquad (4.35)$$

Rearrangement of Equation (4.35) allows γ to be isolated as

$$\gamma = \frac{F_{zc}}{\pi r} + \frac{nr}{l}\gamma_{gb} \qquad (4.36)$$

Equation (4.36) allows calculation of γ by measurement of F_{zc} if γ_{gb} is known. Alternatively, both γ and γ_{gb} may be determined from zero-creep measurements on wires of different radii. The following rearrangement of Equation (4.36) shows that a plot of $F_{zc}/(\pi r)$ versus r will have a slope of $-(n/l)\gamma_{gb}$ and an intercept of γ:

$$\frac{F_{zc}}{\pi r} = \gamma - \frac{n}{l}\gamma_{gb} r \qquad (4.37)$$

Measurements from a single radius may be used to measure both energies if the dihedral angle for the intersection of the grain boundary with the surface of the wire is also measured. Substitution of Equation (4.22) into Equation (4.37) yields

$$\frac{F_{zc}}{\pi r} = \gamma - \frac{2n}{l}\gamma r \cos\left(\frac{\theta_1}{2}\right) = \gamma\left[1 - \frac{2n}{l} r \cos\left(\frac{\theta_1}{2}\right)\right] \qquad (4.38)$$

Thus measurement of F_{zc} and θ_1 allows direct calculation of γ, which may be substituted back into Equation (4.22) to determine γ_{gb}.

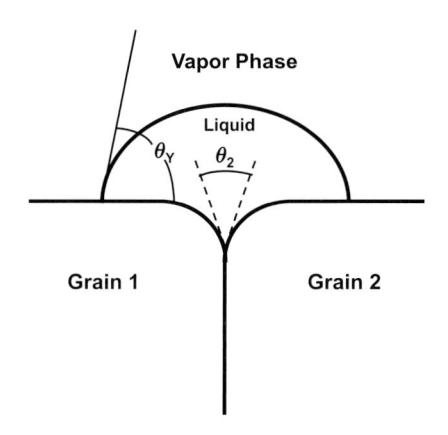

Figure 4.21 A schematic diagram of a grain boundary intersecting a free surface under a liquid droplet.

4.4.2 The multiphase-equilibrium (MPE) technique

Eustathopoulos et al. [15] describe a modification of the grain-boundary-grooving technique for determining surface energies for high-melting-point materials. Figure 4.21 shows a schematic diagram of a grain boundary intersecting a free surface under a liquid droplet of an inert liquid. From this diagram

$$\cos\theta_Y = \frac{\gamma_{SV} - \gamma_{SL}}{\gamma_{LV}} \quad (4.39)$$

and

$$\cos\left(\frac{\theta_2}{2}\right) = \frac{\gamma_{gb}}{2\gamma_{SL}} \quad (4.40)$$

where θ_2 is the dihedral formed under the liquid. If the dihedral angle θ_1 between the grain boundary and free surface is also measured in the absence of the liquid, Figure 4.15, we have

$$\cos\left(\frac{\theta_1}{2}\right) = \frac{\gamma_{gb}}{2\gamma_{SV}} \quad (4.41)$$

Equations (4.39), (4.40) and (4.41) constitute a system of three equations in the three unknowns γ_{SV}, γ_{SL} and γ_{gb} if γ_{LV} is known.

Table 4.4 *Selected values of high-angle grain-boundary energies*

Solid	T (°C)	γ (mJ/m^2)	Reference
Cu	1,030	450	[11]
Cu	925	625	[14]
Cu–20 at% Au	850	430	[14]
Cu–40 at% Au	850	390	[14]
Cu–60 at% Au	850	310	[14]
Cu–80 at% Au	850	320	[14]
Au	1,000	378	[14]
Ni	1,060	866	[14]
Ag	950	375	[14]
Si	1,200	500	[8]
Al$_2$O$_3$	1,200	728[a]	[15]

[a] Results calculated from γ_{SV} and θ_1 from the MPE technique in Reference [15].

4.4.3 Selected values of high-angle grain-boundary energies

Selected values of the surface energies of solids were presented in Table 2.2. Table 4.4 presents selected values of the energies of high-angle grain boundaries.

4.5 Summary

In this chapter the consequences of the orientation-dependent surface energies for crystalline solids have been described. The effects on equilibrium crystal shape and the thermodynamics of grain-boundary behavior and faceting have been used as examples. Two common techniques for measuring γ for solid surfaces, namely zero creep and multiphase equilibrium, have been described.

4.6 References

[1] W. W. Mullins, Solid surface morphologies governed by capillarity. In *Metals Surfaces; Structures, Energetics, Kinetics*, ed. W. D. Robertson and N. A. Gjostein (Metals Park, OH: ASM, 1963), Chapter 2, pp. 17–66.

[2] G. Wulff, Zur Frage der Geschwindigkeit des Wachstums und der Auflösung der Kristallflachen [On the question of speed of growth and dissolution of crystal surfaces], *Z. Kristall. Mineral.*, **34** (1901), 449–530.

[3] B. D. Cullity and S. R. Stock, *Elements of X-ray Diffraction*, 3rd edn. (Upper Saddle River, NJ: Prentice Hall, 2001), Chapter 2.

[4] M. McLean, Determination of the surface energy of copper as a function of crystallographic orientation and temperature, *Acta Metall.*, **19** (1971), 387–393.

[5] C. Herring, Some theorems on the free energies of crystal surfaces, *Phys. Rev.*, **82** (1951), 87–93.

[6] J. Weertman and J. R. Weertman, *Elementary Dislocation Theory* (Oxford: Oxford University Press, 1992).

[7] J. M. Howe, *Interfaces in Materials* (New York: John Wiley & Sons, 1997), Chapters 12 and 13.

[8] A. Otsuki, Energies of (001) twist grain boundaries in silicon, *Acta Mater.*, **49** (2001), 1737–1745.

[9] A. Kelly and G. W. Groves, *Crystallography and Crystal Defects* (Reading, MA, Addison-Wesley, 1970), Chapter 10.

[10] D. A. Porter and K. E. Easterling, *Phase Transformations in Metals and Alloys* (London: Chapman & Hall, 1992), Chapter 3.

[11] M. McLean, Grain boundary energy of copper at 1030 °C, *J. Mater. Sci.*, **8** (1973), 571–576.

[12] L. E. Murr, *Interfacial Phenomena in Metals and Alloys* (Reading, MA: Addison-Wesley Publishing Co., 1975), Chapter 2.

[13] J. L. Meijering, Usefulness of a $1/\gamma$ plot in the theory of thermal etching, *Acta Metall.*, **11** (1963), 847–849.

[14] L. E. Murr, *Interfacial Phenomena in Metals and Alloys* (Reading, MA: Addison-Wesley Publishing Co., 1975), Chapter 3.

4.7 Study problems

1. Use data in Table 4.3 to estimate the dihedral angles for the intersections of both a high-angle grain boundary and an incoherent twin boundary with the surface of Cu if it is annealed at high temperature in vacuum.

2. A fiber of paraffin with a radius of 2.0 mm, which is hanging vertically, is observed to slowly become shorter. It is found that hanging a weight with a mass of 20 mg on the end of the fiber results in a situation such that the fiber neither shortens nor lengthens. Calculate the surface energy of paraffin.

3. When platinum wire with a radius of 0.1 mm is annealed at 1,300 °C, it is observed that the wire develops a bamboo structure, e.g. Figure 4.20(b), with six grain boundaries per mm of length. The grain boundaries are observed to make dihedral angles of approximately 160° with the external surface of the wire. When the wire is hung vertically it is observed that the "zero-creep" condition is established at 1,300 °C when a weight having a mass of 58 mg is attached to the end. Estimate the surface energy and grain boundary energy for Pt at 1,300 °C.

4. Consider the following observations [15].
   (i) When liquid tin is placed on polycrystalline alumina at 1,200 °C it is relatively non-wetting, with $\theta_Y = 121°$.
   (ii) When polycrystalline alumina is heated to 1,200 °C the grain boundaries make average dihedral angles with the surface, as in Figure 4.15, of $\theta_1 = 139°$.
   (iii) The grain boundaries which intersect the Sn/alumina interface, as in Figure 4.21, make dihedral angles of $\theta_2 = 147°$.

If the surface energy of liquid tin is extrapolated to 1,200 °C a value of 470 mJ/m² is estimated. Using this information, estimate the surface energy of alumina, the grain-boundary energy and the interfacial energy between Sn and alumina at 1,200 °C.

5. When liquid Cu is placed on polycrystalline Mo at 1,100 °C the contact angle $\theta_Y = 10°$.
   (a) Assuming that the surface energy of solid Mo is 2,650 mJ/m² at 1,100 °C, calculate the interfacial energy for the Cu (l)/Mo (s) interface.
   (b) Will liquid Cu tend to penetrate the grain boundaries in Mo? Take γ_{gb}^{Mo} to be 700 mJ/m².

6. Consider a small rectangular crystal such as that in Figure 4.4, which has a volume of 2×10^5 nm³. If the surface energies for the various faces are $\gamma_x = 1,200$ mJ/m², $\gamma_y = 1,680$ mJ/m² and $\gamma_z = 1,920$ mJ/m², calculate the length of each side of the crystal when it is in shape equilibrium.

5 Interphase interfaces

In this chapter the structure and thermodynamics of the various interfaces which can form between different phases in solid systems are described. Essentially, all the features described in Chapter 4 for single-phase systems apply, with added complexity being introduced by differences in crystal structure and/or composition of the adjoining phases. Particular emphasis is placed on the effects of surfaces and interfaces on chemical reactions involving thin films.

5.1 Interface classifications

The boundaries between different solid phases may be characterized in terms of their atomic structures and the degree of lattice matching across the interface. Figure 5.1 shows schematic diagrams of the three interface classifications, (a) coherent, (b) semicoherent and (c) incoherent [1].

5.1.1 Coherent interfaces

A coherent interface is formed when the lattices of the two adjacent phases match at the interface as shown in Figure 5.1(a). If the two phases have the same (or similar) crystal structure the lattices can be continuous. An example of this is shown in Figure 5.2, which shows the microstructure in a Ni-base superalloy that is based on the Ni–Al system, Figure 5.3. The two phases are the γ-Ni solid-solution phase which has the fcc (A1) crystal structure and the γ' phase, which has the approximate composition Ni_3Al and has the $L1_2$ crystal structure,

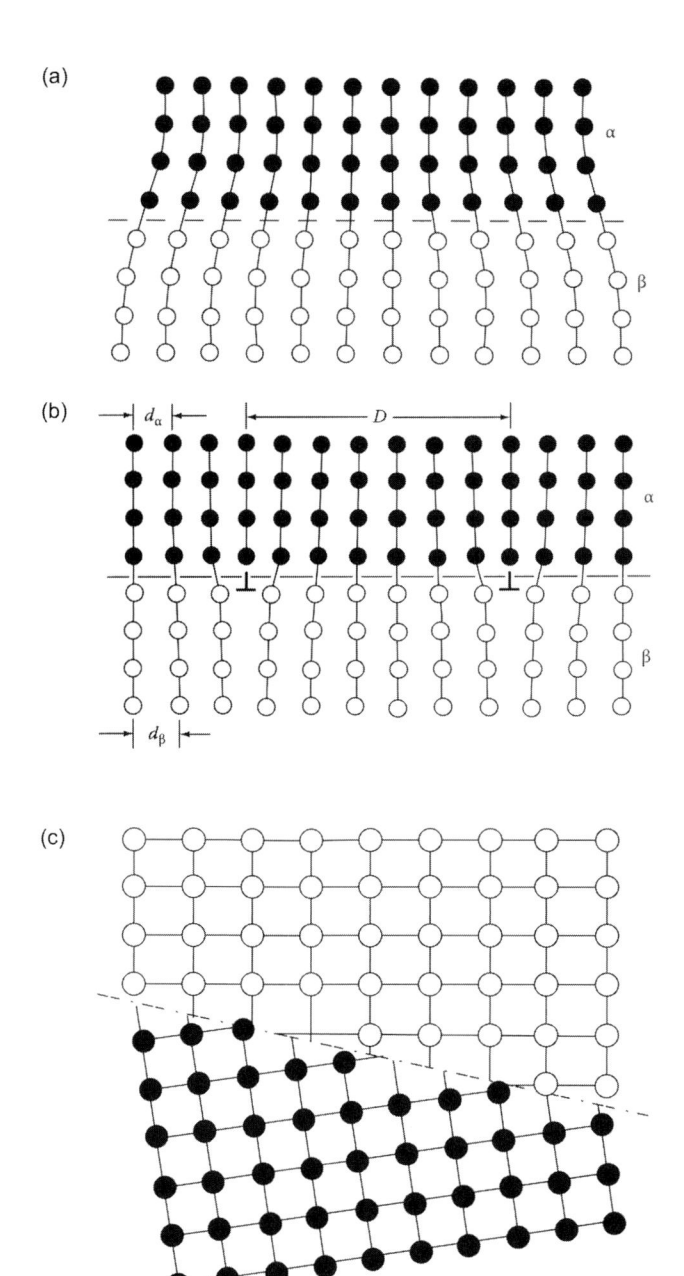

Figure 5.1 Schematic diagrams of the three interface classifications, (a) coherent, (b) semicoherent and (c) incoherent. (These diagrams appeared originally as Figures 3.34, 3.35 and 3.37 in D. A. Porter and K. E. Easterling, *Phase Transformations in Metals and Alloys* (London: Chapman & Hall, 1992) and are reproduced by permission of Springer Science + Business Media.)

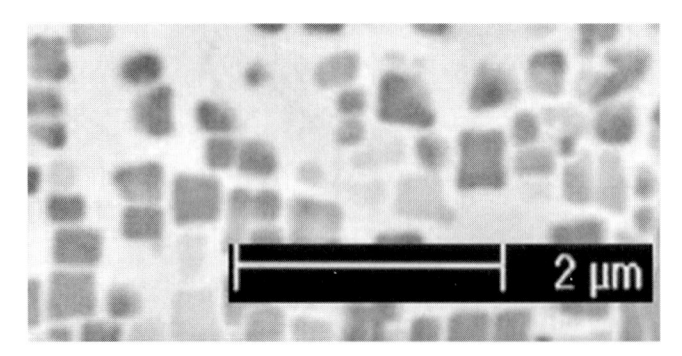

Figure 5.2 The microstructure of a nickel-base superalloy showing cuboidal γ' precipitates in a γ-Ni solid solution.

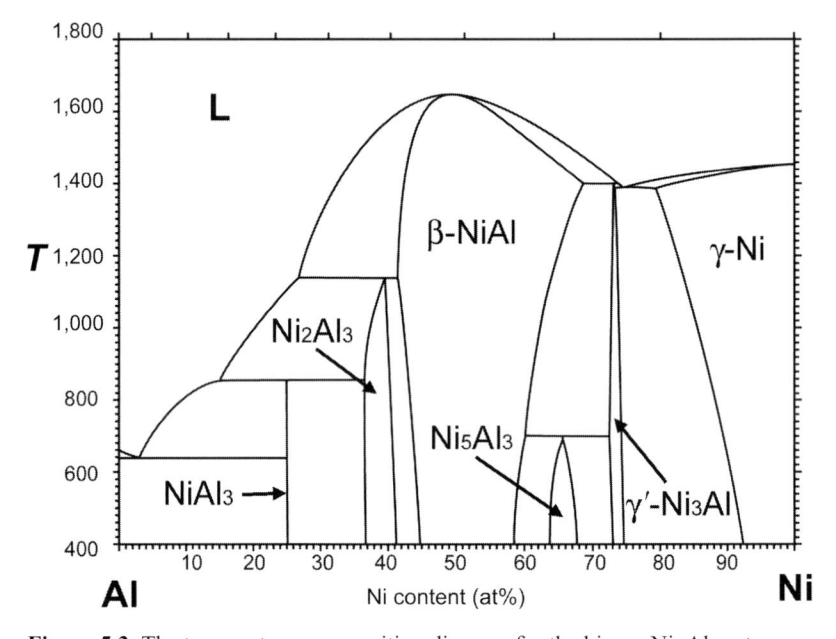

Figure 5.3 The temperature–composition diagram for the binary Ni–Al system.

which is also based on an fcc lattice. The lattice parameters of the two phases are very similar and the {100} planes and the lattices of the two phases are continuous, resulting in an interface with low energy, $\gamma_{\gamma/\gamma'} \approx 20$ mJ/m^2 [2].

The surface energy of a coherent interface results from the atoms being bonded to some wrong neighbors across the interface as can be seen in Figure 5.1(a), i.e. it is a chemical term that depends on the compositions of both phases:

$$\gamma_{coher} = \gamma_{chem} \tag{5.1}$$

If the lattice spacing of the planes which are joined across the interface is different in the two phases, which is also shown in Figure 5.1(a), the two lattices will be elastically strained in order to maintain the coherency. The magnitude of the *coherency strains* and the corresponding coherency stresses, will depend on the misfit between the two lattices,

$$\delta = \frac{d_\beta - d_\alpha}{d_\alpha} \tag{5.2}$$

where d_α and d_β are the interplanar spacings of the planes in the α and β phases which match at the interface. The elastic strain energy resulting from coherency will be proportional to δ^2 [1]. If the misfit is near zero, the precipitates will take the shape of spheres. If the misfit deviates significantly from zero, the precipitates will take the shape that minimizes elastic strain energy. In the case of the superalloy in Figure 5.2, where the misfit

$$\delta = \frac{d_{\gamma'}^{100} - d_{\gamma}^{100}}{d_{\gamma}^{100}} \tag{5.3}$$

is negative, the γ' phase takes the shape of cubes with $\{100\}$ faces.

Figure 5.4 shows the same superalloy after annealing at 1,100 °C for 20 hours showing *coarsening* of the γ' precipitates. This process in which the precipitates grow to minimize the total interfacial free energy of the system will be discussed in Chapter 6. It is introduced here to illustrate another effect of the elastic strain, namely the tendency for *directional coarsening*. Not only have the precipitates grown, but also they have started to line up parallel to $\langle 100 \rangle$ directions to reduce the

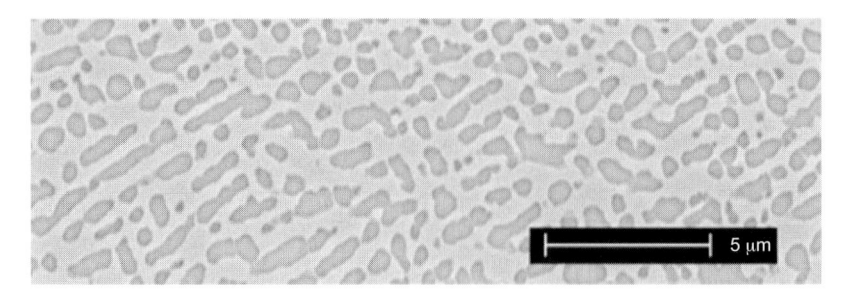

Figure 5.4 The microstructure of a nickel-base superalloy after annealing at 1,100 °C for 20 h showing alignment of the γ' precipitates.

total strain energy and, in some cases, have coalesced. If the alloy is stressed during the anneal, this alignment and coalescence will be more pronounced, leading to a phenomenon known as *rafting*. For the case of alloys with a negative misfit the precipitates will align perpendicular to the stress axis if the loading is tensile and parallel to it if the loading is compressive. If the misfit is positive the alignment directions will be reversed. The process of rafting is described in detail in Reference [3].

5.1.2 Semicoherent interfaces

If the misfit is too large it will be energetically favorable for the interface to contain *misfit dislocations* to take up some of the misfit. This will result in a semicoherent interface, which is shown schematically in Figure 5.1(b). Most of the misfit in one dimension can be accommodated if the dislocations are separated by a distance D given by

$$D = \frac{d_\beta}{\delta} \quad (5.4)$$

Normally an interface will have misfit in two dimensions, which will result in two, non-parallel sets of misfit dislocations. A semicoherent interface may be considered to be somewhat analogous to the low-angle grain boundaries described in Chapter 4.

The interfacial energy of a semicoherent interface will contain two contributions, namely a chemical term associated with the bonding across the interface and a second term arising from the energies of the misfit dislocations:

$$\gamma_{semicoher} = \gamma_{chem} + \gamma_{\perp} \tag{5.5}$$

Thus, semicoherent interfaces will generally have higher interfacial energies but lower elastic strain energies than coherent interfaces. In the case of a coherent second phase, which is distributed in a matrix, as in Figure 5.2, the elastic strain energy will be proportional to the volume of the second phase. Therefore, for a given δ, the interfaces may be coherent when the second-phase particles are small, but may lose coherence as the particles grow. The important phenomenon of coherency loss is well described in Reference [1]. A similar situation exists with the epitaxial growth of crystalline films on a single-crystalline substrate when there is some degree of lattice mismatch between the substrate and film. When the film is thin the interface will be completely coherent (as in Figure 5.1(a)) and as the film thickens and the elastic strain energy increases the interface will become semicoherent (as in Figure 5.1(b)).

5.1.3 Incoherent interfaces

If the lattice misfit is very large, Equation (5.4) indicates that the dislocation spacing will be very small, at which point the boundary structure will be highly disordered, resulting in an *incoherent interface*, which is shown schematically in Figure 5.1(c).

5.1.4 Interface mobility

The subject of interface mobility will not be treated here. However, it should be commented that the mobility of interfaces varies markedly

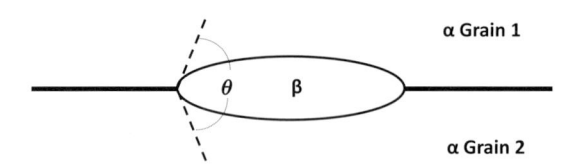

Figure 5.5 A schematic diagram of a two-phase microstructure in which the second phase β has formed on the grain boundaries in the α matrix.

depending on whether they are coherent, semicoherent or coherent. This subject and its applicability to phase transformations are covered in detail in Chapters 3 and 5 of Reference [1] and in Reference [4].

5.2 Interaction of second phases with grain boundaries

If a second phase is present on a grain boundary in a solid the morphology will be influenced by the magnitude of the grain-boundary and interfacial energies. Figure 5.5 shows such a second phase for which the α/β interfaces are incoherent. A free-energy balance at the triple point where the grain boundary and two α/β interfaces meet allows the dihedral angle to be related to the magnitude of the corresponding grain-boundary and interfacial energies,

$$\cos\left(\frac{\theta}{2}\right) = \frac{\gamma_{gb}}{2\gamma_{\alpha\beta}} \qquad (5.6)$$

(This is analogous to the expression for the dihedral angle for a grain-boundary groove in Equation (4.22).)

Figure 5.6 is a transmission electron micrograph showing a TiO_2 precipitate that has formed on the grain boundary of an internally oxidized Cu–Ti alloy [5]. Measurement of the dihedral angle of numerous precipitates yielded values of γ_{Cu/TiO_2} ranging from 400 to 700 mJ/m² when γ_{gb} was approximated as 450 mJ/m² in Equation (5.6). The reason for

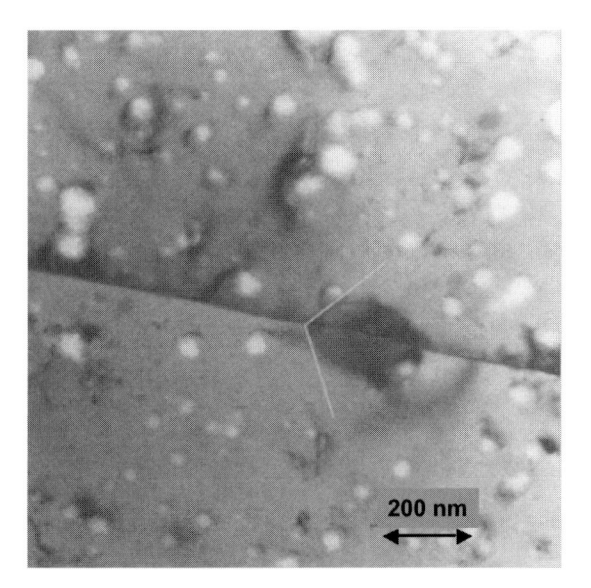

Figure 5.6 A transmission electron micrograph showing a TiO$_2$ precipitate that has formed on a grain boundary of an internally oxidized Cu–Ti alloy.

the considerable spread in values is believed to result from some of the interfaces having some degree of coherency when one or both grains had a favorable orientation to the precipitate. Figure 5.7 is a transmission electron micrograph of a TiO$_2$ precipitate that has been removed from an internally oxidized Cu–Ti alloy in a carbon extraction replica. The crystal has faceted surfaces, suggesting that the Cu/TiO$_2$ interfaces with both grains were semicoherent. Also visible in this micrograph is a groove where a twin boundary in the rutile crystal intersected a Cu/TiO$_2$ interface.

5.3 Thin-film formation

The contribution of surface energies can be significant in systems involving thin films. There are two basic ways in which thin films form

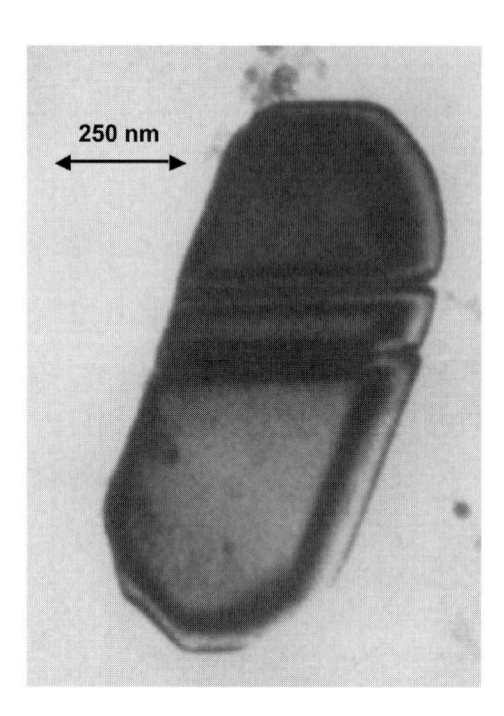

Figure 5.7 A transmission electron micrograph of a TiO$_2$ precipitate that has been removed from an internally oxidized Cu–Ti alloy in a carbon extraction replica.

on a substrate exposed to a gas phase. One is by chemical reaction of a component from the gas phase with one or more components in the substrate. The other is by deposition of one or more components from the gas phase onto the substrate. An example of each of these is presented in Figure 5.8. Figure 5.8(a) illustrates the reaction of oxygen from a gas phase with a metal M to form a film of MO by the reaction

$$M(s) + \tfrac{1}{2}O_2(g) = MO(s) \tag{5.7}$$

Figure 5.8(b) shows the formation of a solid film of metal M_1 on a substrate M_2 by physical vapor deposition (e.g. sputtering, evaporation, etc.) of M_1. Both of these processes allow demonstration of important thermodynamic points.

```
(a)      O₂(g)              (b)      M₁(vapor)
    ─────────────              ─────────────
         MO(s)                       M₁(s)
    ─────────────              ─────────────
         M(s)                        M₂(s)
```

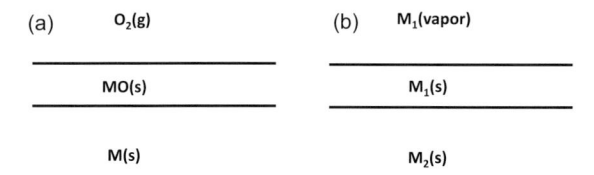

Figure 5.8 Schematic diagrams of two ways in which films form on solid surfaces. (a) Formation of a compound layer by chemical reaction between the atmosphere and the substrate, which is illustrated here for the oxidation of a metal. (b) Formation of a surface layer by chemical vapor deposition, which is illustrated here for the condensation of M_1 vapor onto a M_2 substrate.

5.3.1 Growth of thin oxide films

5.3.1.1 Oxidation of copper

McLean and Hondros [6] studied the effect of oxygen on the surface free energy of copper at oxygen partial pressures below the dissociation pressure of Cu_2O at 1,200 K. (The significance of this type of measurement will be discussed in Chapter 7.) As they increased the oxygen partial pressure from very low values they observed a continuous decrease in γ_{SV}. However, at an oxygen partial pressure of 10^{-16} atm, which is well below the dissociation pressure of Cu_2O, they observed discontinuities in the apparent γ_{SV}, which indicated that a surface film was forming. The formation of Cu_2O occurs by the reaction

$$Cu(s) + \tfrac{1}{2}O_2(g) = Cu_2O(s) \tag{5.8}$$

The free-energy change for the formation of Cu_2O via reaction (5.6) for bulk phases is

$$\Delta G = \Delta G^\circ + RT \ln \left(\frac{a_{Cu_2O}}{a_{Cu}^2 p_{O_2}^{1/2}} \right) \tag{5.9}$$

For the oxidation of pure Cu, the activities of Cu and Cu_2O may be taken as unity and setting $\Delta G = 0$ yields the dissociation pressure

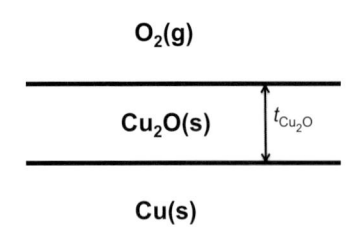

Figure 5.9 A schematic diagram of a Cu_2O film forming on Cu.

of Cu_2O,

$$p_{O_2}^{eq} = \exp\left(\frac{2\Delta G°}{RT}\right) \quad (5.10)$$

At 1,200 K $p_{O_2}^{eq}$ is approximately 10^{-8} atm. (See, for example, Figure 1.13.) However, if the oxide is forming as a thin layer then the surface energies must be included in this calculation. The formation of the oxide results in the replacement of a Cu/gas surface by a Cu_2O/gas surface and a Cu/Cu_2O interface, Figure 5.9. Therefore the total free-energy change will be

$$\Delta G = \Delta G° + RT \ln\left(\frac{a_{Cu_2O}}{a_{Cu}^2 p_{O_2}^{1/2}}\right)$$
$$+ \left[\gamma_{Cu_2O/gas} + \gamma_{Cu/Cu_2O} - \gamma_{Cu/gas}\right]\frac{V_{Cu_2O}}{t_{Cu_2O}} \quad (5.11)$$

where V_{Cu_2O} is the molar volume of Cu_2O and t_{Cu_2O} is the oxide thickness. As can be seen from Table 4.3, $\gamma_{Cu/gas} \approx 1,400$ mJ/m². McLean and Hondros estimated the sum $\gamma_{Cu_2O/gas} + \gamma_{Cu/Cu_2O}$ to be approximately 800 mJ/m², i.e. the surface-energy term in Equation (5.11) is negative. This means that ΔG can be negative for values of p_{O_2} that are smaller than $p_{O_2}^{eq}$ as long as the oxide thickness is less than a limiting value. The limiting thickness at which oxide growth will stop may be calculated from Equation (5.11) by setting $\Delta G = 0$ and $a_{Cu} = a_{Cu_2O} = 1$. This thickness is plotted as a function of p_{O_2} in Figure 5.10. When $p_{O_2} > p_{O_2}^{eq}$ the oxide will continue to grow. Similar observations were made for the oxidation of iron [7]. In most situations involving a new phase, the

5.3 THIN-FILM FORMATION

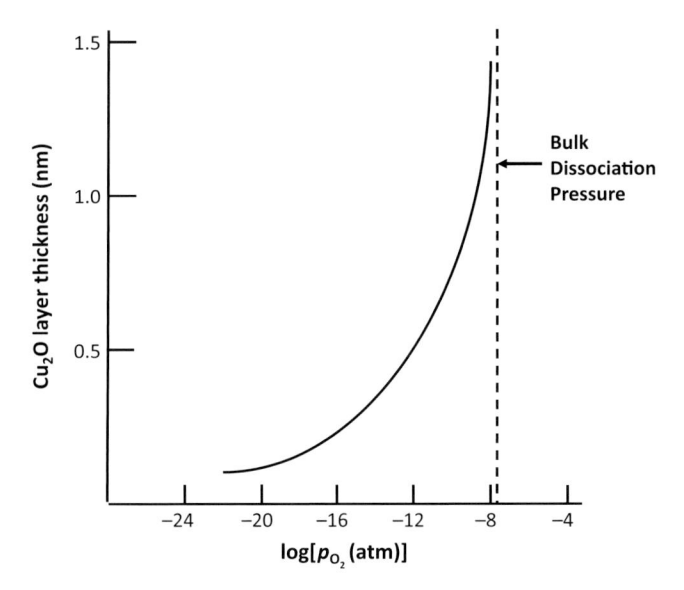

Figure 5.10 A plot showing the effect of oxygen partial pressures, which are below the bulk dissociation pressure of Cu_2O, on the limiting thickness of the oxide layer (after Reference [4]).

surface-energy term is considered to provide a barrier to *nucleation* of the new phase, as will be discussed in Chapter 6. However, in cases such as the oxidation of Cu and Fe the surface-energy term contributes to the thermodynamic driving force for formation of the new phase in a special case of *heterogeneous nucleation*.

5.3.1.2 Oxidation of silicon

The oxidation of Si by oxygen or water vapor to produce an electrically resistive SiO_2 layer is an important part of the fabrication of semiconductor devices [8]. The silica layer that forms is observed to be vitreous rather than crystalline when thin and becomes crystalline as it thickens, as shown schematically in Figure 5.11. The existence of the vitreous oxide is generally explained in terms of the kinetics of oxide growth. However, Jeurgens *et al.* [9] have recently reanalyzed this case and propose that

140 INTERPHASE INTERFACES

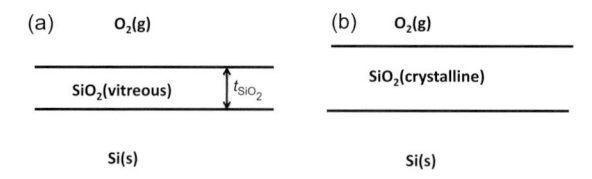

Figure 5.11 A schematic diagram of silica forming on elemental Si. (a) Vitreous silica when the oxide is thin. (b) Crystalline silica when the oxide is thick.

the vitreous silica is, in fact, the thermodynamically stable phase up to a limiting thickness if the surface and interfacial energies are considered. Jeurgens et al. have presented a detailed analysis, which includes strains in the crystalline oxide. The following is a simpler analysis to illustrate the important features of the problem. Consider the formation of silica by the reaction

$$\text{Si(s)} + \text{O}_2(\text{g}) = \text{SiO}_2(\text{cryst, vitreous}) \tag{5.12}$$

The free-energy difference between the crystalline and vitreous layers will be

$$\Delta G = \Delta G^\circ_{\text{SiO}_2(\text{vitreous})} - \Delta G^\circ_{\text{SiO}_2(\text{cryst})}$$

$$+ \left[\gamma^{\text{vitreous}}_{\text{SiO}_2/\text{gas}} + \gamma^{\text{vitreous}}_{\text{Si}/\text{SiO}_2} - \gamma^{\text{cryst}}_{\text{SiO}_2/\text{gas}} - \gamma^{\text{cryst}}_{\text{Si}/\text{SiO}_2} \right] \frac{V_{\text{SiO}_2}}{t_{\text{SiO}_2}} \tag{5.13}$$

where $\Delta G^\circ_{\text{SiO}_2(\text{vitreous})}$ and $\Delta G^\circ_{\text{SiO}_2(\text{cryst})}$ are the standard free energies of formation of crystalline and vitreous silica, respectively, the γ terms are the energies for the indicated surfaces and interfaces, t_{SiO_2} is the thickness of the silica film and V_{SiO_2} is the molar volume of silica. The difference between the molar volume of vitreous and crystalline silica has been omitted for simplicity in Equation (5.13). Figure 5.12 is a plot of the standard free energies of formation for liquid and crystalline (cristobalite) SiO_2 as functions of temperature. (The data have been taken from Reference [10].) The lines, of course, cross at the melting temperature of SiO_2 (1,996 K) but the entropy differences are small enough that, even at a typical oxidation temperature of 700 K, the difference between the

5.3 THIN-FILM FORMATION 141

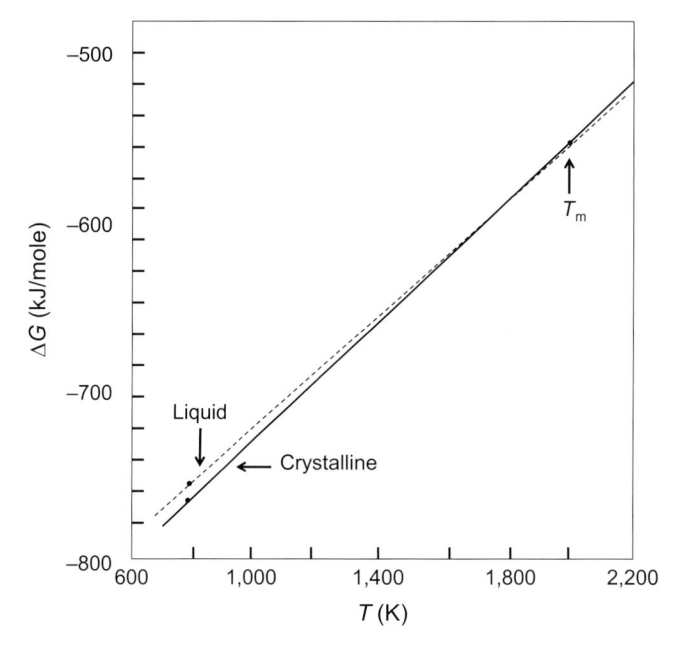

Figure 5.12 A plot of the standard free energies of formation for liquid and crystalline (cristobalite) SiO_2 as a function of temperature.

two free energies is only 3 kJ. Jeurgens et al. [9] have calculated the surface and interfacial energies such that the term in square brackets is negative with a value of approximately 2 J/m². Taking the molar volume as approximately 10^{-5} m³/mole and setting $\Delta G = 0$ in Equation (5.13) yields a critical thickness of approximately 10 nm. The more detailed analysis [9] indicates that vitreous silica is the thermodynamically stable phase up to thicknesses of approximately 40 nm.

Calculations for other oxides for which the standard free-energy difference between amorphous and crystalline oxides is large yield correspondingly smaller critical thicknesses [9]. For example, for MgO on Mg the value is on the order of a few tenths of a nanometer, i.e. less than a lattice parameter, such that vitreous MgO would not be expected to form.

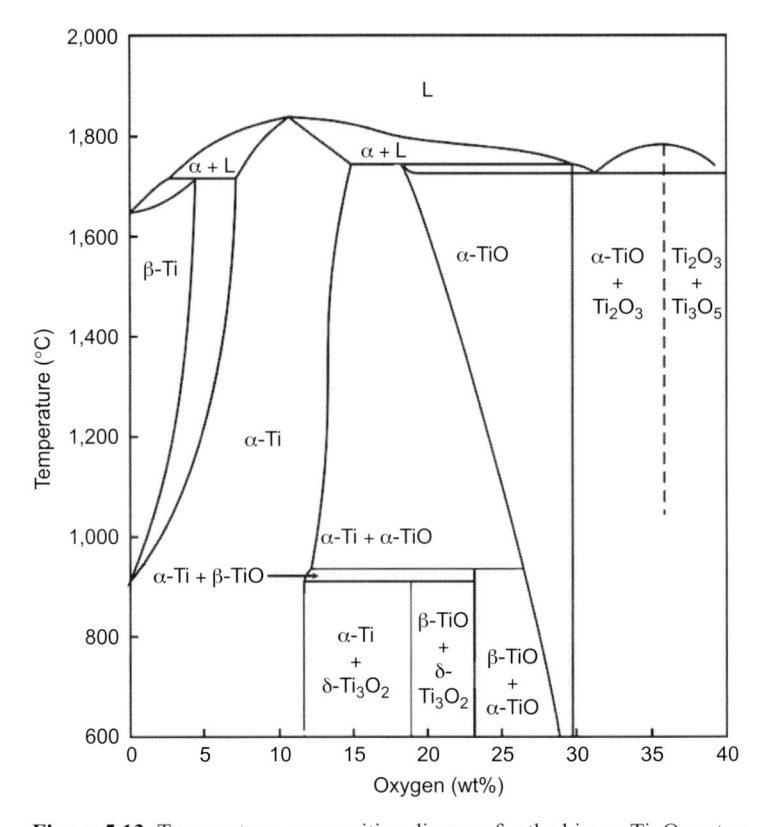

Figure 5.13 Temperature–composition diagram for the binary Ti–O system.

5.3.1.3 Oxidation of titanium

The oxidation of Ti indicates another aspect of the influence of surfaces and interfaces on thermodynamics, namely the absence of phases that would be expected to be stable on the basis of bulk thermodynamics. Figure 5.13 is the temperature–composition diagram for the binary Ti–O system. Figure 5.14(a) is a schematic diagram of an oxide film growing on Ti with the sequence of layers predicted by thermodynamics, i.e. Figure 5.13. However, when Ti is oxidized the oxide layer usually consists only of TiO_2, as shown in Figure 5.14(b). Table 5.1 shows typical values of equilibrium oxygen partial pressures at 1,100 K. Consider a relatively

Table 5.1 *Equilibria in the Ti–O system*

Equilibrium	Oxygen partial pressure (atm)
Ti/TiO	10^{-42}
TiO/Ti$_2$O$_3$	10^{-33}
Ti$_2$O$_3$/Ti$_3$O$_5$	10^{-28}
Ti$_3$O$_5$/TiO$_2$	10^{-25}

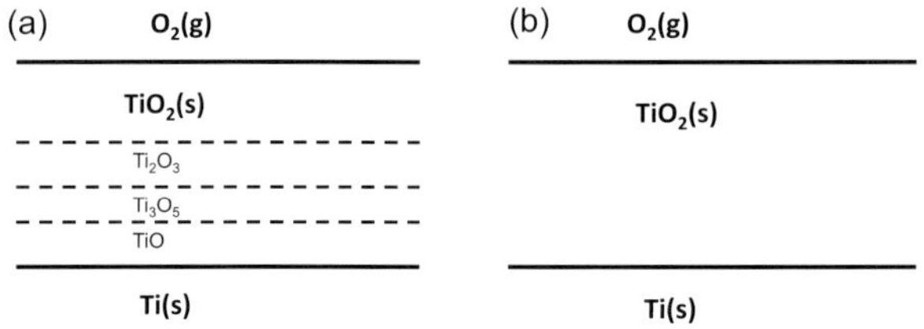

(a) O$_2$(g)

TiO$_2$(s)
Ti$_2$O$_3$
Ti$_3$O$_5$
TiO

Ti(s)

(b) O$_2$(g)

TiO$_2$(s)

Ti(s)

Figure 5.14 A schematic diagram of an oxide film growing on Ti. (a) The sequence of layers predicted by thermodynamics. (b) The usually observed oxide consisting only of TiO$_2$.

thick oxide layer of 10 μm. If one assumes an oxygen gradient across the oxide layer that is essentially linear from 1 atm to 10^{-42} atm, the spacing between the positions for the various oxygen partial pressures in Table 5.1 would be on the orders of fractions of a nanometer. Thus, even if the gradient were not linear, the thickness over which the lower oxides have the lowest free energy would not be great enough to supply the energy of the interface which would need to be created.

5.3.2 Formation of metal films by evaporation

5.3.2.1 Formation of thin films of Ag by vapor deposition onto Cu

The formation of a layer of a new phase on a solid substrate also affects the surface and interface stresses in the system. For example, the deposition of a layer of silver onto copper results in the consumption

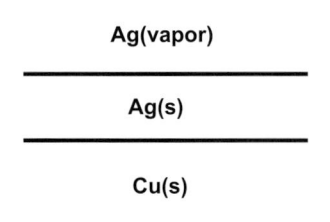

Figure 5.15 A schematic diagram of the formation of a film by vapor deposition of Ag onto Cu.

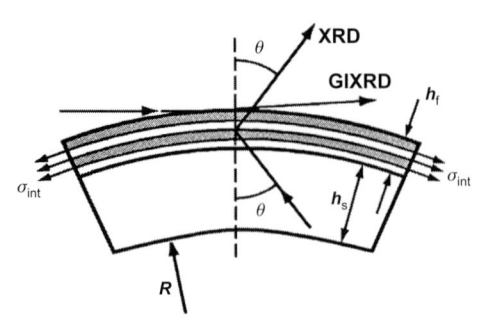

Figure 5.16 A schematic diagram of the measurement of interfacial stress in a multilayered film. GIXRD, grazing-incidence X-ray diffraction. (From [11], reprinted with permission of Elsevier.)

of a Cu/gas surface and the production of an Ag/gas surface and a Cu/Ag interface, Figure 5.15. The deposition of a layer of copper onto the silver layer would create a morphology consisting of a Cu/gas surface and two Cu/Ag interfaces. In most cases the measurement of the interface stress (σ_{int}) is not possible because of other contributions to stresses in the films from sources such as thermal-expansion mismatch. (This situation will be described in Chapter 8.) However, in the case of thin-film deposition, it is possible to use techniques such as that shown in Figure 5.16 [11] to measure σ by combining bending-curvature measurement and stress measurement by X-ray diffraction (XRD). If the thickness of the multilayered film is small relative to the thickness of the substrate, the force per unit width exerted by the film on the substrate is related to the radius of curvature R by

$$\frac{F}{w} = \frac{1}{6}\frac{E_S}{1-\nu_S}h_S^2\frac{1}{R} \tag{5.14}$$

where h_S, E_S and ν_S are the thickness, Young's modulus and Poisson's ratio for the substrate, respectively. The force in Equation (5.14) is the result of two contributions: the average stress in the film (intrinsic and thermal) and the interface stresses,

$$\frac{F}{w} = \langle \sigma \rangle + N\sigma_{\text{int}} \quad (5.15)$$

where $\langle \sigma \rangle$ represents the average film stress, which may be determined from measurement of the in-plane lattice parameters by XRD [12], and N is the number of interfaces in the multilayered film. Measurements on (111)-textured Ag/Cu multilayers yielded values of $\sigma_{\text{int}} = -3.19 \pm 0.43$ N/m $= -3.19 \pm 0.43$ J/m^2. Compressive interface stresses were also measured for nickel/silver multilayered films [11]. Note that there is only a single value for σ_{int} because the tensor quantity σ is isotropic for planes/interfaces with three-fold or higher symmetry [13].

Cammarata [14] has reviewed the subject of interface stress and presents theoretical values for several metal/metal interfaces, which were calculated using embedded-atom methods. The calculated interfacial stresses for (111)/(111) interfaces both for Ag/Cu and for Ag/Ni were tensile. The calculated value for Ag/Cu was 0.32 J/m^2, while the experimental value is -3.19 J/m^2. The cause of this discrepancy is not known but the experimental value should at worst be correct with regard to sign.

5.4 Summary

In this chapter it has been shown that interphase interfaces in crystalline materials may be characterized by their structure as coherent, semicoherent or incoherent. The nature of the interface affects the morphology of second phases and the way they interact with grain boundaries. It has also been shown that the relative values of the surface and interfacial energies in a layered system can have a significant effect on the stability of second phases. Finally, the interface stress in thin-film systems has been briefly discussed.

5.5 References

[1] D. A. Porter and K. E. Easterling, *Phase Transformations in Metals and Alloys* (London: Chapman & Hall, 1992), Chapter 3.

[2] L. E. Murr, *Interfacial Phenomena in Metals and Alloys* (Reading, MA: Addison-Wesley Publishing Co., 1975), Chapter 3.

[3] M. Fahrmann, W. Hermann, E. Fahrmann, A. Boegli, T. M. Pollock and H. G. Sockel, Determination of matrix and precipitate elastic constants in (γ–γ') Ni-base model alloys and their relevance to rafting, *Mater. Sci. Eng., A*, **260** (1999), 212–221.

[4] J. M. Howe, *Interfaces in Materials* (New York: John Wiley & Sons, 1997), Chapter 10.

[5] S. Wood, D. Adamonis, A. Guha, W. A. Soffa and G. H. Meier, Internal oxidation of dilute Cu–Ti alloys, *Metall. Trans.*, **6A** (1975), 1793–1800.

[6] M. McLean and E. D. Hondros, Interfacial energies and chemical compound formation, *J. Mater. Sci.*, **8** (1973), 349–351.

[7] E. D. Hondros, The effect of adsorbed oxygen on the surface energy of bcc iron, *Acta Metall.*, **16** (1968), 1377–1380.

[8] F. P. Fehlner, *Low Temperature Oxidation: The Role of Vitreous Oxides* (New York: Wiley-Interscience, 1989).

[9] L. P. H. Jeurgens, Z. Wang and E. J. Mittemeijer, Thermodynamics of reactions and phase transformations at interfaces and surfaces, *Int. J. Mater. Res.*, **100** (2009), 1281–1307.

[10] JANAF Thermochemical Tables, *J. Phys. Chem. Ref. Data*, Vol. 14, Suppl. 1 (1985).

[11] F. Spaepen, Interfaces and stresses in thin films, *Acta Mater.*, **48** (2000), 31–42.

[12] I. C. Noyan and J. B. Cohen, *Residual Stresses* (Berlin: Springer-Verlag, 1987).

[13] R. Shuttleworth, The surface tension of solids, *Proc. Phys. Soc. (London)*, **A63** (1950), 444–457.

[14] R. C. Cammarata, Surface and interface stress effects in thin films, *Prog. Surf. Sci.*, **46** (1994), 1–37.

5.6 Study problems

1. The TiO_2 particles on the grain boundaries of Cu (Figure 5.6) formed dihedral angles with the grain boundaries of approximately 132° at 900 °C. What would the expected dihedral angle be for the intersection of grain boundaries with the free surface for these alloys at the same temperature? Take γ_{SV} for Cu from Table 2.2 and assume γ for the Cu/TiO_2 interface to be 550 mJ/m².
2. Consider the formation of a film of Cu_2O on metallic copper by reaction with an atmosphere in which the oxygen partial pressure is low. Given the data below, calculate the following quantities:
   (a) the dissociation pressure of Cu_2O at 1,300 K; and
   (b) the thickness of the Cu_2O films which could form on Cu at 1,300 K if the oxygen partial pressure in the atmosphere is 10^{-8} atm.

   For Cu_2O, $\Delta G° = -162{,}200 + 69.24T$ J/mole Cu_2O, $\gamma_{Cu/gas} \approx 1{,}630$ mJ/m² and $\gamma_{Cu_2O/gas} + \gamma_{Cu/Cu_2O} \approx 800$ mJ/m² $V_{Cu_2O} \approx 24$ cm³/mole.

6 Curved surfaces

In this chapter the effects of surface curvature on the thermodynamics of various types of systems are described. Some examples are as follows.

1. The pressure beneath a curved surface – the Laplace equation.
2. The effect of curvature on the chemical potential – the Gibbs–Thomson equation.
3. Phase equilibria, e.g. the effect of size on the vapor pressure of a liquid droplet, the effect of size on the melting point of a solid, shifts of phase boundaries on phase diagrams when one of the phases has small dimensions.
4. Applications, nucleation, coarsening of fine precipitates, grain growth in polycrystalline materials, etc.

Particular emphasis is placed on the effects of size on phase equilibria and phase diagrams.

6.1 Derivation of the Laplace equation

This treatment follows that in Reference [1], except that it is not restricted to fluid phases by inserting the *surface energy*. Rather the *surface stress* is used, which makes the derivation valid for solid and fluid phases. Consider the curved surface in Figure 6.1. This could be the surface of a bubble in a liquid or solid. The pressures inside and outside the bubble will be different because of the action of the surface stress. In the simplest case of a bubble within a liquid, where the surface stress is equal to the surface energy and is always positive, the pressure inside

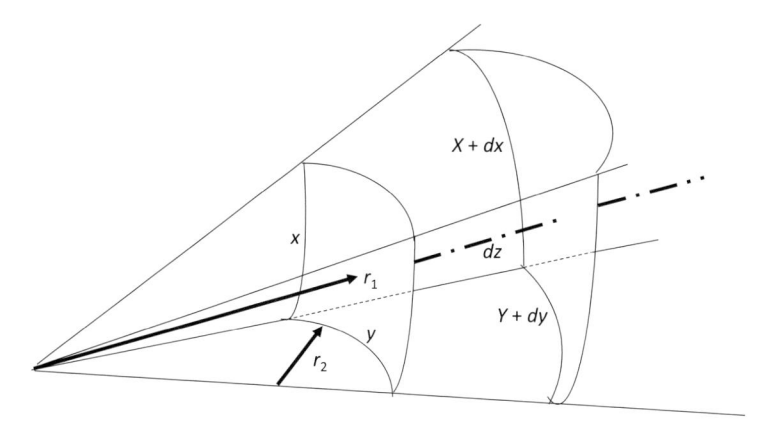

Figure 6.1 Displacement of the surface of a bubble, which allows the calculation of the pressure difference across the bubble wall. (From [1], reprinted with permission of John Wiley & Sons.)

the bubble must be greater than that outside to prevent the bubble from collapsing. In the case of a bubble in a solid the interpretation is not as simple, since the surface stress may be positive or negative. In any event, assume that the surface is displaced from its initial position by a distance dz. The segment of the surface with initial area xy will have a final area $(x + dx)(y + dy)$. Therefore, the change in area will be

$$\Delta A = (x + dx)(y + dy) - xy = x\,dy + y\,dx \qquad (6.1)$$

The non P–V work associated with changing (deforming) the area of the surface will be

$$\delta w' = \sigma(x\,dy + y\,dx) \qquad (6.2)$$

where σ refers to either σ_{LV} or σ_{SV}. The first law requires that the surface work is equal to the P–V work associated with the pressure difference across the surface,

$$\delta w = \Delta P\,dV' = \Delta P\,xy\,dz \qquad (6.3)$$

such that

$$\sigma(x\,dy + y\,dx) = \Delta P\,xy\,dz \tag{6.4}$$

The relationships of dz to dx and dy may be established from the geometry in Figure 6.1. From similar triangles,

$$\frac{x + dx}{r_1 + dz} = \frac{x}{r_1} \tag{6.5}$$

and

$$\frac{y + dy}{r_2 + dz} = \frac{y}{r_2} \tag{6.6}$$

Multiplying both sides of Equation (6.5) by $r_1 + dz$ yields

$$x + dx = \frac{xr_1}{r_1} + \frac{x\,dz}{r_1} \tag{6.7}$$

which reduces to

$$dx = \frac{x\,dz}{r_1} \tag{6.8}$$

Similar multiplication through Equation (6.6) by $r_2 + dz$ results in

$$dy = \frac{y\,dz}{r_2} \tag{6.9}$$

Substitution of Equations (6.8) and (6.9) into the left-hand side of Equation (6.4) yields

$$\sigma\left(\frac{xy\,dz}{r_1} + \frac{xy\,dz}{r_2}\right) = \Delta P\,xy\,dz \tag{6.10}$$

which, upon division through by $xy\,dz$, becomes the general form of the Laplace equation,

$$\Delta P = \sigma\left(\frac{1}{r_1} + \frac{1}{r_2}\right) \tag{6.11}$$

For the often-encountered special case of spherical geometry, $r_1 = r_2 = r$, the Laplace equation takes the simple form

$$\Delta P = \frac{2\sigma}{r} \qquad (6.12)$$

It is emphasized that the proper surface quantity to be used in Equations (6.11) and (6.12) is σ, not γ. It is only for the special case of fluid systems, where $\sigma = \gamma$, that the surface energy may be used in the Laplace equation and

$$\Delta P = \frac{2\gamma}{r} \qquad (6.13)$$

6.1.1 Techniques that use the Laplace equation to measure surface energy

In Chapter 2 the capillary-rise technique for measuring γ_{LV} was described. A number of other techniques, which are based on the Laplace equation, may also be used to measure this quantity. These are briefly described here.

6.1.1.1 The sessile-drop technique

In Chapter 3 the determination of the contact angle using a sessile drop was discussed. However, this technique may also be used to measure γ_{LV} from the size and shape of the drop. For a reasonably large drop the profile will be somewhat flattened under the influence of gravity. (See Figure 3.3.) Therefore the radii of curvature will vary along the drop, depending on the hydrostatic pressure and the pressure defined by Equation (6.11). A detailed analysis of the drop geometry allows calculation of $\sigma_{LV} = \gamma_{LV}$. Detailed discussions of this procedure are provided in References [2] and [3].

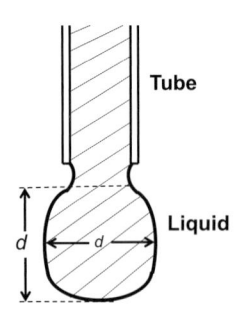

Figure 6.2 A schematic diagram of the pendant-drop technique for measuring surface energies of liquids.

6.1.1.2 The pendant-drop technique

A drop protruding from the end of a capillary tube, Figure 6.2, will be elongated by gravity. Again there will be a balance between the hydrostatic pressure and the pressure determined by the Laplace equation. Analysis of the drop geometry allows γ_{LV} to be determined from

$$\gamma_{SV} = \chi g \rho d^2 \tag{6.14}$$

where ρ is the liquid density and χ is a factor that is tabulated for different drop shapes. Detailed descriptions of this technique are available in References [1] and [2].

6.1.1.3 The maximum-bubble-pressure technique

If a capillary tube is inserted to a depth h below the surface of a liquid the hydrostatic pressure will be $\rho g h$. If an inert gas is blown through the tube the pressure in the resultant bubble will be given from Equation (6.13) as

$$P = \frac{2\gamma_{LV}}{r} + \rho g h \tag{6.15}$$

The maximum pressure will obtain for the minimum value of r, which corresponds to the point when the bubble is hemispheric in shape and r will be the radius of the tube. Measurement of the maximum value of P

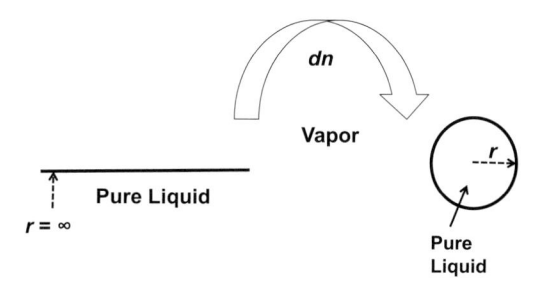

Figure 6.3 Transfer of matter from a pure liquid with a planar surface to a droplet of the same liquid with radius r.

allows calculation of γ_{LV}. A discussion of the methods of measurement and potential sources of error is presented in Reference [1].

6.2 The effect of curvature on the chemical potential

In Chapter 1 the importance of the chemical potential in calculations of phase equilibria was described. It is now necessary to describe the changes to chemical potentials which occur when the phase of interest is small in size so that contributions from surfaces/interfaces become important.

Consider the reversible transfer of matter shown schematically in Figure 6.3, which occurs isothermally. A small amount of matter is transported through the vapor phase from a planar liquid to a spherical liquid droplet. If the chemical potential in the bulk liquid is $\mu(\infty)$ the chemical potential in the droplet may be determined by adding the free-energy change associated with the transfer. Presume that one mole of atoms is transferred. (In order that the radius of the droplet does not change, we may assume that the transfer is occurring to a very large number of droplets, all of the same size.) According to Equation (1.31) the change will be $V_L^\circ \Delta P$, where V_L° is the molar volume of the liquid. It is assumed the liquid is incompressible so the molar volume is the

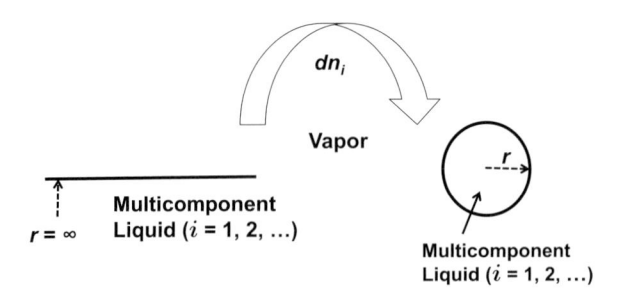

Figure 6.4 Transfer of component *i* from a multicomponent liquid with a planar surface to a droplet of the same liquid with radius *r*.

same despite the higher pressure in the droplet. We then have

$$\mu(r) = \mu(\infty) + V_L^\circ \Delta P \qquad (6.16)$$

The change in pressure will be given by the Laplace equation so that

$$\mu(r) - \mu(\infty) = \frac{2\gamma}{r} V_L^\circ \qquad (6.17)$$

This relationship is known as the *Gibbs–Thomson* equation. Thus, the chemical potential in the droplet is greater than that in the bulk liquid and the difference increases as the size of the droplet decreases.

A similar relationship may be derived for each component in a multicomponent liquid using the transfer shown schematically in Figure 6.4 in which component *i* is transferred. In this case the free-energy change for the transfer is $\bar{V}_i \Delta P$ and application of the Laplace equation yields a Gibbs–Thomson equation for each component in the liquid:

$$\mu_i(r) - \mu_i(\infty) = \frac{2\gamma}{r} \bar{V}_i \qquad (6.18)$$

where $\bar{V}_i$ is the partial molar volume of component *i*.

The same transfer as described by Figure 6.3 could also be envisioned from a pure bulk solid to a small solid particle. However, in this case the Laplace equation must be written as Equation (6.12) and the

6.2 EFFECT OF CURVATURE ON CHEMICAL POTENTIAL

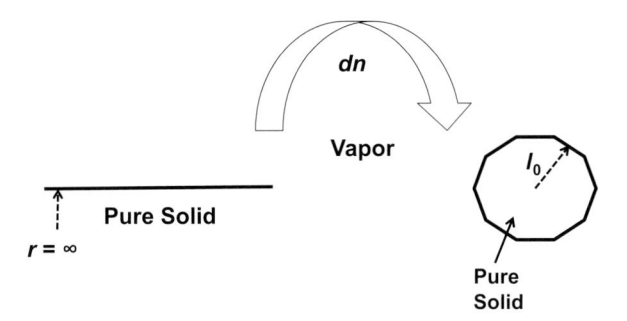

Figure 6.5 Transfer of matter from a pure solid with a planar surface to a small faceted crystal of the same solid.

Gibbs–Thomson equation becomes

$$\mu(r) - \mu(\infty) = \frac{2\sigma}{r} V_S^\circ \tag{6.19}$$

In the solid case, the change in chemical potential is determined by the surface stress. Similarly, if the transfer in Figure 6.4 involved a multicomponent solid solution,

$$\mu_i(r) - \mu_i(\infty) = \frac{2\sigma}{r} \bar{V}_i \tag{6.20}$$

Johnson [4] has used a similar approach to derive a *generalized Gibbs–Thomson equation* with which to describe the chemical potential in a crystalline solid. Consideration of the transport between a bulk solid and a faceted crystal, as shown schematically in Figure 6.5, leads to

$$\mu(l_0) - \mu(\infty) = \frac{2\gamma_0}{l_0} V_S^\circ \tag{6.21}$$

which may be compared with Equation (6.19). The chemical potential is uniform throughout the crystal since the ratio γ/l is the same for each crystal facet because of the Wulff theorem. However, this derivation would seem to be applicable only to a special case where the surface stress for the solid may be replaced by the surface energy.

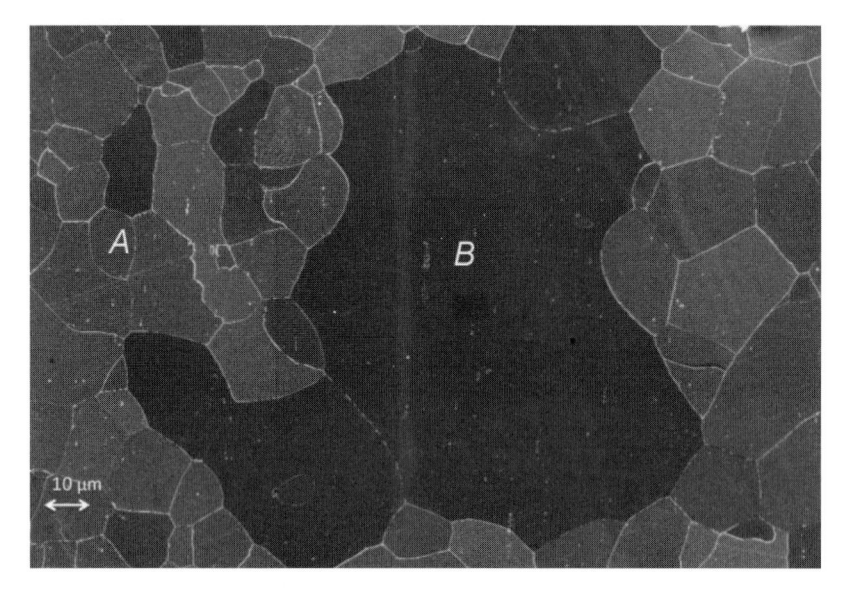

Figure 6.6 A scanning electron micrograph of the microstructure of a Si-steel during grain growth. (Micrograph provided courtesy of Dr. M. Hua.)

The above analyses were restricted to transport across a free surface. However, the Gibbs–Thomson equation is applicable to grain boundaries as well as free surfaces. The application to grain growth in polycrystalline materials is described in the next section.

6.2.1 Grain growth

The microstructure of a Si-steel that is undergoing grain growth is presented in Figure 6.6. If there is a distribution of grain sizes in such a polycrystalline material there will be a tendency for large grains to grow and small grains to shrink. This can be seen from the simple schematic diagrams in Figure 6.7. As discussed in Section 4.2.3, if the grain-boundary energy is independent of orientation, a six-sided grain will be at equilibrium with straight boundaries and dihedral angles of 120°. Small grains will have fewer than six sides and the boundaries will be

Figure 6.7 A schematic diagram showing grain-boundary curvature for grains with various numbers of sides.

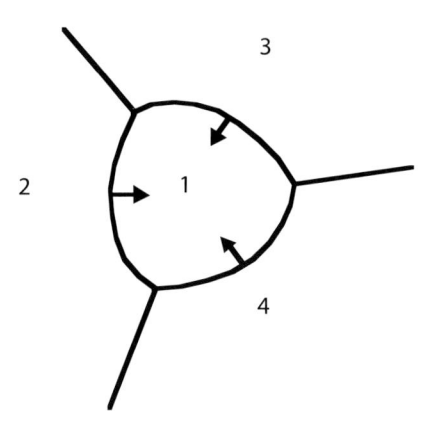

Figure 6.8 A schematic diagram showing the direction of motion of the grain boundaries for a three-sided grain.

concave toward the grain center, while larger grains will have more than six sides and the boundaries will be concave away from the grain center. Curvature of the boundaries will result in a chemical-potential difference across the boundary. Consider the boundaries of the grain in Figure 6.8. The chemical-potential difference across the boundary between grains 1 and 2 will be, using Equation (6.19),

$$\mu_1 - \mu_2 = 2\gamma_{\text{gb}} V_S^\circ \left[\frac{1}{r_1} - \frac{1}{r_2}\right] \quad (6.22)$$

However, r_1 is positive and r_2 is negative, i.e. $r_2 = -r_1$ so that

$$\mu_1 - \mu_2 = \frac{4\gamma_{\text{gb}} V_S^\circ}{r_1} \quad (6.23)$$

Therefore there is a driving force for the transport of matter from grain 1 to grain 2 such that grain 1 will shrink and grain 2 will grow.

A simple kinetic analysis [5] gives the change in the average grain size with time as

$$\bar{D}^2 = \bar{D}_0^2 + \kappa t \quad (6.24)$$

where $\bar{D}$ is the mean grain "diameter" at time t, $\bar{D}_0$ is the initial grain size and κ is a rate constant that is proportional to the product of the grain-boundary energy and the grain-boundary mobility.

Figure 6.6 also illustrates another phenomenon, *abnormal grain growth*. In some instances an alloy contains an array of small particles of radius r (too small to resolve in Figure 6.6), which impede the grain-boundary motion. The pinning force is inversely proportional to r [5]. When the particles coarsen, according to a capillarity-driven process to be described in Section 6.5.2, the grain boundaries are able to "break away" and move rapidly. This process is stochastic in nature and the first grains to break away grow rapidly, as can be seen in Figure 6.6.

6.3 Phase equilibria in one-component systems

6.3.1 The relation between μ_S and μ_L (or μ_V)

Consider a small spherical piece of an isotropic solid in a large quantity of fluid (liquid or vapor), Figure 6.9. Assume a small amount of material is reversibly transferred from the fluid to the solid. The changes in the

6.3 PHASE EQUILIBRIA IN ONE-COMPONENT SYSTEMS

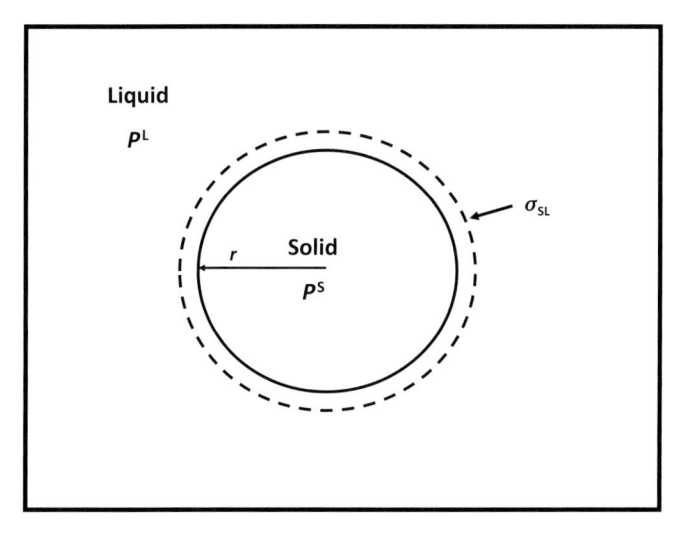

Figure 6.9 A schematic diagram of a spherical solid in equilibrium with the liquid in a unary system.

Helmholtz free energy for the two phases will be

$$dF_S = -P^S dV^S - S^S dT + \mu_S dn^S \tag{6.25}$$

and

$$dF_L = -P^L dV^L - S^L dT + \mu_L dn^L \tag{6.26}$$

Assuming that the entire system is at a constant temperature and the transfer does not change the system volume,

$$dF_{\text{Total}} = 0 = dF_S + dF_L + \gamma\, dA \tag{6.27}$$

Substituting the Laplace equation for the pressure in the solid yields

$$-\left(P^L + \frac{2\sigma}{r}\right)dV^S + \mu_S\, dn^S + P^L\, dV^S - \mu_L\, dn^S + \gamma\, dA = 0 \tag{6.28}$$

Equation (6.28) may be simplified to

$$-\frac{2\sigma}{r}dV^S + (\mu_S - \mu_L)dn^S + \gamma\, dA = 0 \tag{6.29}$$

where dA and dV^S are related for the spherical geometry as follows:

$$A = 4\pi r^2, \qquad dA = 8\pi r \, dr \qquad (6.30)$$

$$V^S = \tfrac{4}{3}\pi r^3, \qquad dV^S = 4\pi r^2 \, dr \qquad (6.31)$$

and

$$dA = \frac{2 \, dV^S}{r} \qquad (6.32)$$

Replacement of dA in Equation (6.29) using Equation (6.32) results in

$$\mu_S - \mu_L = \frac{2(\sigma - \gamma) \, dV^S}{r \, dn^S} = \frac{2(\sigma - \gamma) V_S^\circ}{r} \qquad (6.33)$$

The same equation was derived by Gibbs [6] using the internal energy. This equation clearly applies to equilibrium between any two phases. For the special cases where $\sigma = \gamma$, e.g. both phases are fluids, Equation (6.33) reduces to the case obtained for planar interfaces, i.e. the chemical potentials are equal. Both γ and σ appear in Equation (6.33). Cahn [7] has pointed out that those processes which change the surface (interface) area cause γ to be introduced and those which change the state of strain (pressure) cause σ to be introduced.

6.3.2 The vapor pressure of a pure liquid

Equation (6.33) may be used to relate the vapor pressure of a liquid droplet to its size. Consider Figure 6.10 for the case of a spherical droplet. In this case the right-hand side of Equation (6.33) is zero and

$$\mu^V = \mu^L \qquad (6.34)$$

For an infinitesimally small change in the radius of the droplet

$$d\mu^V = d\mu^L \qquad (6.35)$$

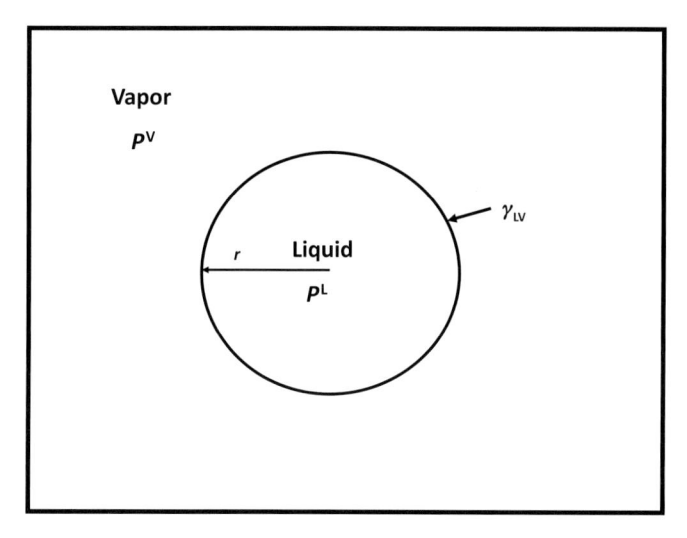

Figure 6.10 A schematic diagram of a liquid droplet in equilibrium with its vapor in a unary system.

At constant temperature Equation (6.35) becomes

$$V^V dP^V = V^L dP^L \tag{6.36}$$

and the pressure difference across the liquid/vapor interface is given by the form of the Laplace equation given as

$$P^L - P^V = \frac{2\gamma}{r} \tag{6.37}$$

For the small change in radius

$$dP^L - dP^V = d\left(\frac{2\gamma}{r}\right) \tag{6.38}$$

Substituting for dP^L from Equation (6.38) into Equation (6.36) yields

$$\frac{V^V - V^L}{V^L} dP^V = d\left(\frac{2\gamma}{r}\right) \tag{6.39}$$

The molar volume of the vapor will be much greater than that of the liquid and, if it is assumed that the vapor is an ideal gas, the difference in molar volumes becomes

$$V^V - V^L \approx V^V = \frac{RT}{P^V} \qquad (6.40)$$

Substitution of Equation (6.40) into Equation (6.39) yields an expression that may be integrated as follows:

$$\frac{RT}{V^L} \int_{P(\infty)}^{P(R)} \frac{dP^V}{P^V} = \int_{1/R=0}^{1/R} d\left(\frac{2\gamma}{r}\right) \qquad (6.41)$$

Upon integration this yields the well-known *Kelvin equation* which relates the vapor pressure of a liquid droplet of radius r to that for the same liquid in bulk:

$$\ln\left[\frac{P(r)}{P(\infty)}\right] = \frac{V^L}{RT}\frac{2\gamma}{r} \qquad (6.42)$$

6.3.3 The vapor pressure of an isotropic solid particle

Consider the spherical solid particle in Figure 6.11, which is assumed to have isotropic surface properties. In this case calculation of the vapor pressure requires the use of Equation (6.33) and, for an infinitesimal change in radius,

$$d\mu_V + d\left[\frac{2(\sigma - \gamma)}{r}V_S^\circ\right] = d\mu_S \qquad (6.43)$$

At constant temperature Equation (6.43) becomes

$$V^V dP^V + d\left[\frac{2(\sigma - \gamma)}{r}V_S^\circ\right] = V_S^\circ dP^S \qquad (6.44)$$

In this case the Laplace equation must be inserted using the form in Equation (6.12),

$$dP^S = dP^V + d\left(\frac{2\sigma}{r}\right) \qquad (6.45)$$

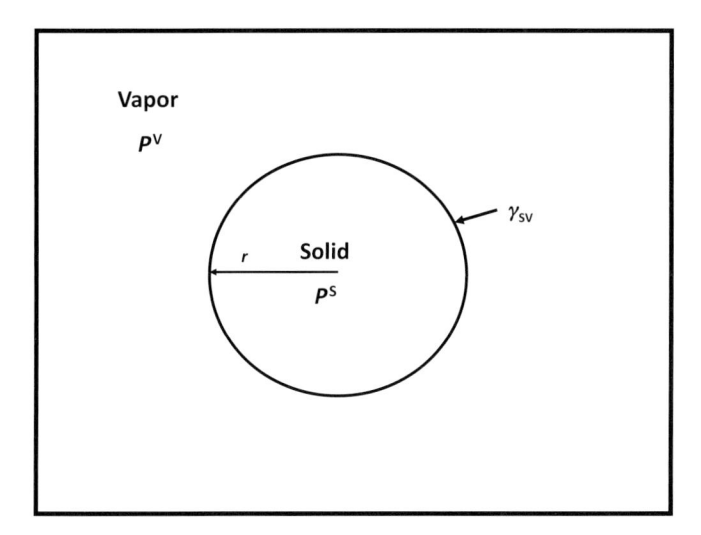

Figure 6.11 A schematic diagram of an isotropic solid particle in equilibrium with its vapor in a unary system.

and Equation (6.44) becomes

$$V^V dP^V + d\left[\frac{2(\sigma - \gamma)}{r} V_S^\circ\right] = V_S^\circ dP^V + d\left(\frac{2\sigma V_S^\circ}{r}\right) \quad (6.46)$$

which may be simplified to

$$\frac{V^V - V_S^\circ}{V_S^\circ} dP^V = d\left(\frac{2\gamma}{r}\right) \quad (6.47)$$

Note that the surface stress is eliminated. If assumptions are made in a similar manner to those in Equation (6.40) for the liquid, the molar volume difference between solid and vapor becomes

$$V^V - V_S^\circ \approx V^V = \frac{RT}{P^V} \quad (6.48)$$

Substitution of Equation (6.48) into Equation (6.47) yields an expression that may be integrated as follows:

$$\frac{RT}{V_S^\circ} \int_{P(\infty)}^{P(R)} \frac{dP^V}{P^V} = \int_{1/R=0}^{1/R} d\left(\frac{2\gamma}{r}\right) \quad (6.49)$$

which integrates to

$$\ln\left[\frac{P(r)}{P(\infty)}\right] = \frac{V_S^\circ}{RT}\frac{2\gamma}{r} \tag{6.50}$$

Thus the Kelvin equation also describes the vapor pressure of an isotropic solid. The description of the vapor pressure of a small crystal becomes a little more complex, but the appropriate surface function to describe the equilibrium is still γ.

6.3.4 The melting point of a one-component solid

The melting point of an isotropic solid is lowered as the particle radius is decreased. Assume a spherical solid. Equation (6.33) yields

$$d\mu_S = d\mu_L + d\left[\frac{2(\sigma - \gamma)V_S^\circ}{r}\right] \tag{6.51}$$

which, upon substitution of the combined first and second laws, results in

$$-S^S\,dT + V_S^\circ\,dP^S = -S^L\,dT + V_L^\circ\,dP^L + d\left[\frac{2(\sigma - \gamma)V_S^\circ}{r}\right] \tag{6.52}$$

and substitution of the Laplace equation for dP^S yields

$$-S^S\,dT + V_S^\circ\left[dP^L + d\left(\frac{2\sigma}{r}\right)\right] = -S^L\,dT + V_L^\circ\,dP^L$$

$$+ d\left[\frac{2(\sigma - \gamma)V_S^\circ}{r}\right] \tag{6.53}$$

If P^L is assumed to remain constant, Equation (6.53) simpifies to

$$(S^L - S^S)dT = -V_S^\circ\,d\left(\frac{2\gamma}{r}\right) \tag{6.54}$$

Integration of Equation (6.54) from $r = \infty$ (for a flat interface) to $r = r$ may be performed as follows:

$$\int_{T_m(\infty)}^{T_m(r)} dT = -\frac{2\gamma V_S^\circ}{\Delta S_m^\circ}\int_0^{1/r} d\left(\frac{1}{r}\right) \tag{6.55}$$

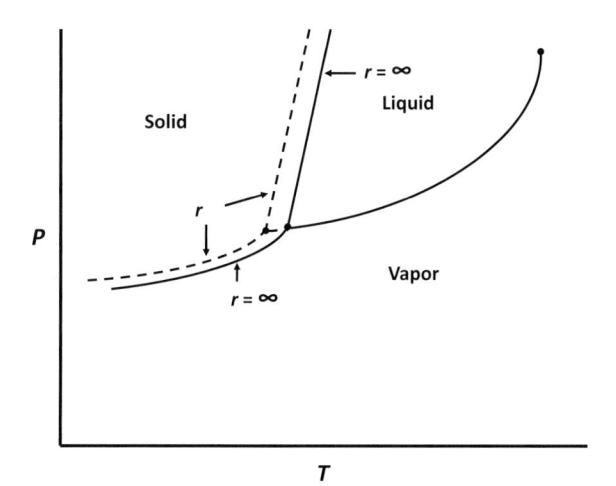

Figure 6.12 The P–T diagram for a unary system, showing the effect of particle size of the solid on the solid–liquid and solid–vapor equilibrium.

The melting point increases with decreasing radius of the solid according to

$$T_m(r) - T_m(\infty) = -\frac{2\gamma_{SL} V_S^\circ}{r \Delta S_m^\circ} \approx -\frac{2\gamma_{SL} T_m(\infty)}{r \Delta H_m^\circ} \quad (6.56)$$

This effect is illustrated in the P–T diagram in Figure 6.12, which also shows the effect of radius on the vapor pressure of the solid (Equation (6.50)).

Note that, for the important equations describing phase equilibria for unary systems when one of the phases is solid, Equations (6.50) and (6.56), the surface stress cancels out from the derivations and the only surface parameter necessary is the surface energy. This is not always the case when multicomponent systems are considered [7].

6.4 Nucleation

Generally, the formation of a new phase by nucleation will require extra energy to form the new surface/interface with the parent phase. Consider

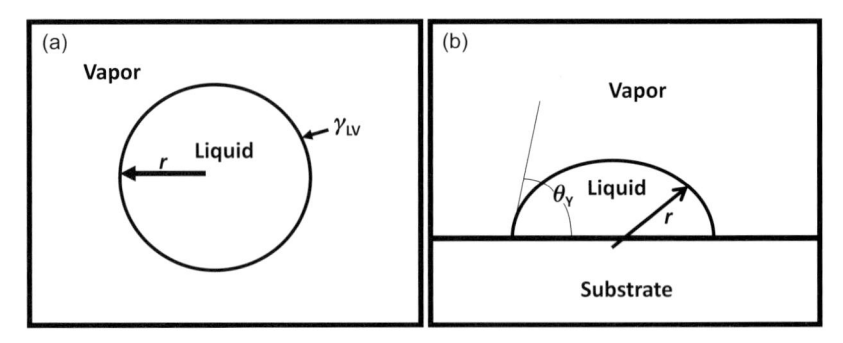

Figure 6.13 Schematic diagrams showing the nucleation of liquid from a vapor for the cases of (a) homogeneous nucleation and (b) heterogeneous nucleation.

the classical nucleation problem of forming a spherical nucleus of liquid from the vapor.

6.4.1 Homogeneous nucleation

The formation of a spherical liquid nucleus in the "bulk" vapor is shown schematically in Figure 6.13(a). The free energy of formation of a nucleus of radius r is

$$\Delta G = 4\pi r^2 \gamma + \frac{4}{3}\pi r^3 \Delta G_V \tag{6.57}$$

where γ is the surface energy of the liquid and ΔG_V is the free-energy change per unit volume of liquid formed during the reaction. The surface and volume terms, along with ΔG, are plotted as a function of r in Figure 6.14. Nuclei with radii greater than r^* will tend to grow spontaneously. At r^*

$$\frac{d\Delta G}{dr} = 8\pi r^* \gamma + 4\pi r^{*2} \Delta G_V = 0 \tag{6.58}$$

which may be solved for r^* as

$$r^* = -\frac{2\gamma}{\Delta G_V} \tag{6.59}$$

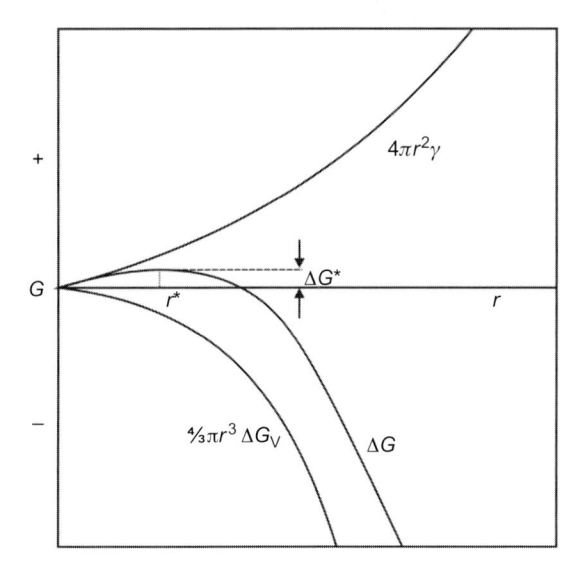

Figure 6.14 The free energy of formation of a nucleus as a function of radius.

Substitution from Equation (6.59) into Equation (6.57) yields ΔG^*, the height of the activation barrier to nucleation:

$$\Delta G^* = 4\pi \frac{4\gamma^3}{\Delta G_V^2} - \frac{4}{3}\pi \frac{8\gamma^3}{\Delta G_V^3}\Delta G_V = \left(16 - \frac{32}{3}\right)\pi \frac{\gamma^3}{\Delta G_V^2} = \frac{16}{3}\pi \frac{\gamma^3}{\Delta G_V^2} \quad (6.60)$$

It is this value, ΔG^*, which enters into the nucleation rate as calculated from absolute reaction-rate theory,

$$J = \omega C^* = \omega C_0 \exp[-\Delta G^*/(RT)] \quad (6.61)$$

where C_0 is the concentration of reactant molecules, C^* is the concentration of critical nuclei, and ω is the frequency with which atoms are added to the critical nucleus. Clearly, since the nucleation rate is proportional to the exponential of a quantity, which in turn depends on the cube of the surface energy, this parameter will have a drastic effect on the nucleation rate, which will increase with decreasing γ.

6.4.2 Heterogeneous nucleation

The nucleation of the same liquid on a substrate is shown schematically in Figure 6.13(b).

In this case the liquid need only form a portion of the sphere in order to achieve the critical radius and the free energy of formation of the nucleus will be multiplied by a shape factor, $S(\theta_Y)$, which will depend on how well the liquid wets the substrate,

$$\Delta G^* = \frac{16}{3}\pi \frac{\gamma^3}{\Delta G_V^2} S(\theta_Y) \tag{6.62}$$

where

$$S(\theta_Y) = \tfrac{1}{4}(2 + \cos\theta_Y)(1 - \cos\theta_Y)^2 \tag{6.63}$$

If the liquid wets poorly, $S(\theta_Y)$ will be near unity. However, in the limiting case of complete wetting, where $\cos\theta_Y$ approaches unity, $S(\theta_Y)$ goes to zero. Thus the rate of heterogeneous nucleation is generally much more rapid than that of homogeneous nucleation, which is rarely observed.

The reader is referred to Chapters 4 and 5 in Reference [5] for a detailed discussion of nucleation in liquid–solid and solid–solid phase transformations.

6.5 Phase equilibria in multicomponent systems

The phase equilibrium in a multicomponent system is altered when one of the phases is finely divided, i.e. has a sharp curvature. In this section the effects of curvature on vapor pressures and solubilities will be discussed.

6.5.1 The vapor pressure of a component over a multicomponent liquid

Consider the case of the liquid–vapor equilibrium shown in Figure 6.15. For each component the chemical potentials are equal,

$$\mu_i^V = \mu_i^L \tag{6.64}$$

6.5 PHASE EQUILIBRIA IN MULTICOMPONENT SYSTEMS

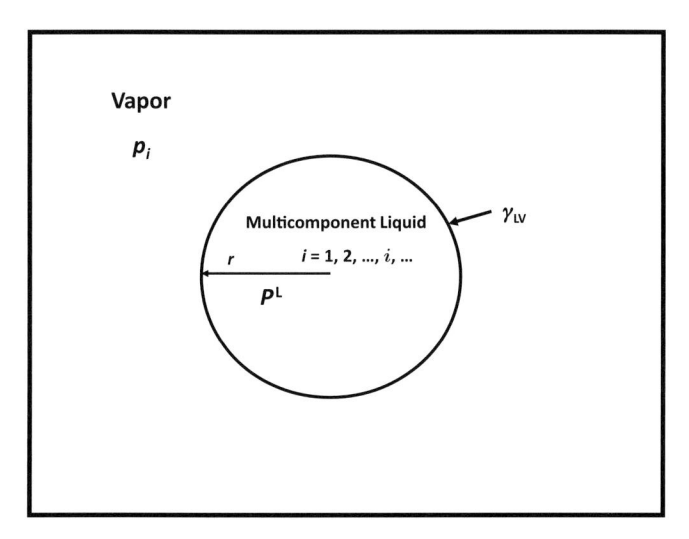

Figure 6.15 A schematic diagram showing a multicomponent liquid in equilibrium with its own vapor.

For an infinitesimally small change in the radius of the droplet,

$$d\mu_i^V = d\mu_i^L \tag{6.65}$$

At constant temperature, insertion of Equation (6.28) for μ_i^L and the approximation in Equation (6.40) for the volume of the vapor species yields

$$\frac{RT}{p_i} dp_i = d\left(\frac{2\gamma}{r}\right) \bar{V}_i \tag{6.66}$$

Integration of Equation (6.66) may be performed as follows:

$$\frac{RT}{\bar{V}_i} \int_{p_i(\infty)}^{p_i(r)} \frac{dp_i}{p_i} = \int_{1/r=0}^{1/r} d\left(\frac{2\gamma}{r}\right) \tag{6.67}$$

Upon integration this yields the logarithm of the ratio of the vapor pressure of component i in the liquid droplet of radius r to that for the same liquid in bulk,

$$\ln\left[\frac{p_i(r)}{p_i(\infty)}\right] = \frac{\bar{V}_i}{RT} \frac{2\gamma}{r} \tag{6.68}$$

170 CURVED SURFACES

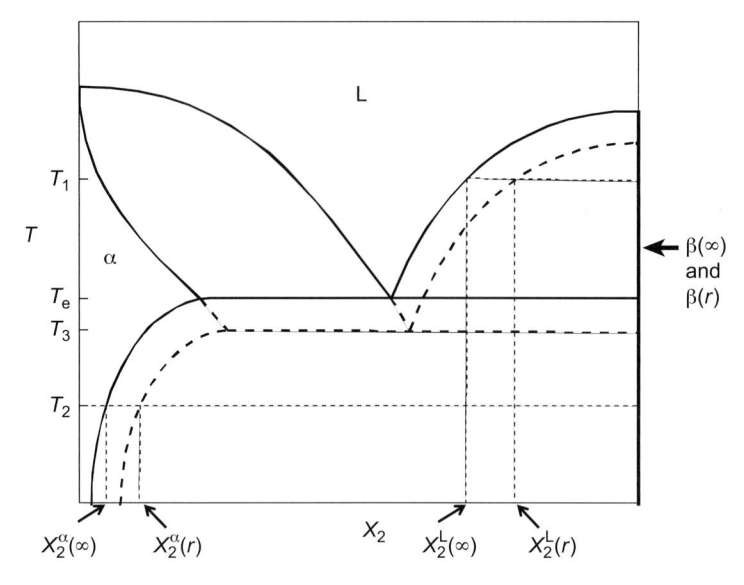

Figure 6.16 A binary temperature–composition diagram with a eutectic reaction, showing how the phase boundary shifts if one phase, β, is present as small spherical particles.

Equation (6.68) is the analog of the Kelvin equation for a multicomponent liquid.

6.5.2 The effect of particle size on solubility

Consider the case of the simple binary T–X diagram in Figure 6.16, for which there is negligible solubility of the α phase or the liquid phase in the β phase, i.e. β consists of virtually pure component 2. The solid curves represent the diagram for all bulk phases. The dashed curves represent the diagram for when the β phase is very finely divided (here taken to consist of small spheres of radius r). The shifts in the phase boundaries are directly analogous to the shifts caused by metastability. (See Figures 1.11 and 1.12.) The free-energy–composition diagrams at

6.5 PHASE EQUILIBRIA IN MULTICOMPONENT SYSTEMS

selected temperatures are presented in Figure 6.17. It may be seen from Figures 6.16 and 6.17 that the following apply.

1. The melting point of β is lowered as the radius becomes smaller. (This is the result described by Equation (6.56) and Figure 6.12.)
2. The solubility of β in the liquid is increased.
3. The solubility of β in α is increased.

The increase in solubility in the liquid may be quantified as follows. Consider the L–β equilibrium at temperature T_1. The chemical potential of component 2 in the liquid is related to its activity by Equation (6.69) for equilibrium with bulk β,

$$\mu_2^L(\infty) = \mu_2^{\circ,L} + RT \ln a_2(\infty) \tag{6.69}$$

The corresponding relation when the liquid is in equilibrium with β spheres of radius r is given by

$$\mu_2^L(r) = \mu_2^{\circ,L} + RT \ln a_2(r) \tag{6.70}$$

The difference between the two chemical potentials is thus given by

$$\mu_2^L(r) - \mu_2^L(\infty) = RT \ln \left(\frac{a_2(r)}{a_2(\infty)} \right) \tag{6.71}$$

Equality of the chemical potentials in coexisting bulk phases is given by

$$\mu_2^L(\infty) = \mu_2^\beta(\infty) = \mu_2^\beta(r) - \frac{2\sigma V_2^\circ}{r} \tag{6.72}$$

Therefore, insertion of Equation (6.72) for $\mu_2^L(\infty)$ into Equation (6.71) yields

$$\mu_2^L(r) - \mu_2^\beta(r) + \frac{2\sigma V_2^\circ}{r} = RT \ln \left(\frac{a_2(r)}{a_2(\infty)} \right) \tag{6.73}$$

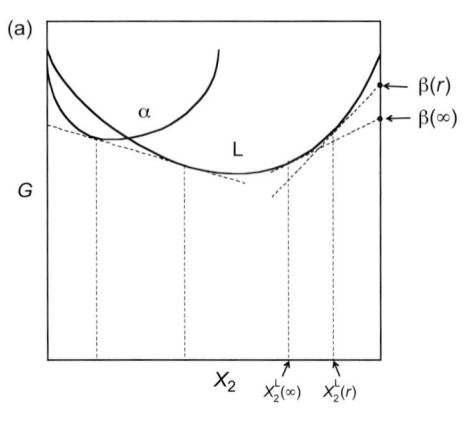

Figure 6.17 Free-energy–composition diagrams corresponding to equilibria at selected temperatures on Figure 6.15: (a) T_1, (b) T_2 and (c) T_3.

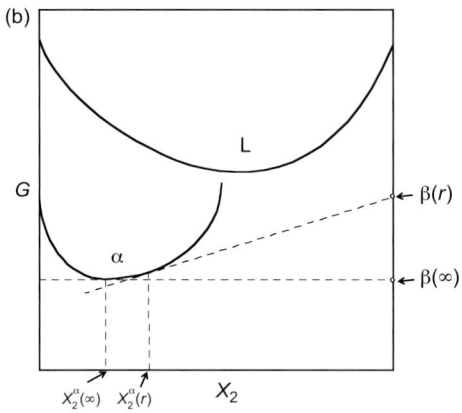

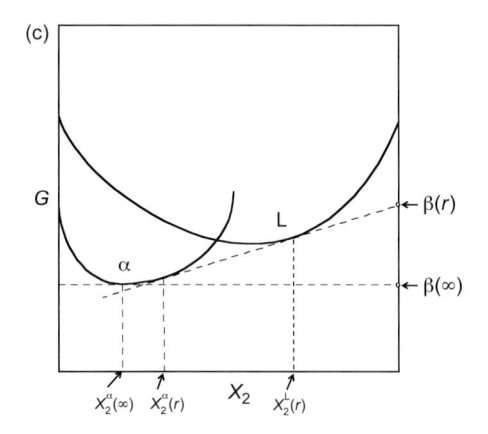

6.5 PHASE EQUILIBRIA IN MULTICOMPONENT SYSTEMS

Also the difference expressed by the first two terms in Equation (6.73) may be replaced by Equation (6.33) to yield

$$-\frac{2(\sigma - \gamma)V_2^\circ}{r} + \frac{2\sigma V_2^\circ}{r} = RT \ln\left(\frac{a_2(r)}{a_2(\infty)}\right) \tag{6.74}$$

which simplifies to

$$\ln\left(\frac{a_2(r)}{a_2(\infty)}\right) = \frac{2\gamma V_2^\circ}{rRT} \tag{6.75}$$

Finally, in the trivial case of the liquid being an ideal solution, the ratio of activities becomes a ratio of concentrations. Also, if the activity coefficient in the liquid is not a strong function of concentration then

$$\ln\left(\frac{C_2(r)}{C_2(\infty)}\right) = \frac{2\gamma V_2^\circ}{rRT} \tag{6.76}$$

Equation (6.76) is known as the *Thomson–Freundlich equation*, which can also be written as

$$C_2(r) = C_2(\infty) \exp\left(\frac{2\gamma V_2^\circ}{rRT}\right) \tag{6.77}$$

For cases where $2\gamma V_2^\circ/(rRT) \ll 1$, Equation (6.77) may be approximated by

$$C_2(r) \approx C_2(\infty)\left(1 + \frac{2\gamma V_2^\circ}{rRT}\right) \tag{6.78}$$

A similar relation will describe the effect of curvature on the solvus curve in Figure 6.16, with the β/L interfacial energy being replaced by the β/α interfacial energy.

6.5.2.1 Comments on the use of free-energy–composition diagrams

In Chapter 1 it was shown that the equality of chemical potential and the common-tangent construction were equivalent ways of describing phase equilibria in multicomponent systems. However, when one or more of the phases is small enough that surface/interface effects become important, equality of chemical potential does not obtain when solid phases are involved. This raises the question of whether the common-tangent construction can still be used. For example, in Figure 6.17 the shift in the chemical potential of B in β (intercepts of the tangents at $X_2 = 1$) is $2\sigma/r$, while the change in solubility of β in liquid, Equation (6.77), was determined by $2\gamma/r$. Therefore, the free-energy–composition diagrams, while useful to qualitatively show the capillarity shifts, must be used with care. Hillert and Agren [8] have presented a treatment in which they redefine the chemical potentials in the solid phase in an attempt to maintain the common-tangent construction. The reader is referred to the original publication for details.

6.5.2.2 The effect of particle size on the solubility of components from solid solution in a liquid

The section on solubility will be completed with a discussion of a slightly more general case where the solid phase β is a binary substitutional solution. Jesser et al. [9] have shown that the equivalent of Equation (6.33) for a multicomponent system is

$$\mu_i^S - \mu_i^L = \frac{2(\sigma - \gamma)\bar{V}_i}{r} \tag{6.79}$$

The relation between activity and chemical potential in the liquid phase will again be

$$\mu_2^L(r) - \mu_2^L(\infty) = RT \ln \left(\frac{a_2^L(r)}{a_2^L(\infty)} \right) \tag{6.80}$$

If the activity coefficients are assumed to vary slowly with composition then

$$\frac{a_2^L(r)}{a_2^L(\infty)} \approx \frac{C_2^L(r)}{C_2^L(\infty)} \tag{6.81}$$

Insertion of Equation (6.81) into Equation (6.80) yields

$$\mu_2^L(r) - \mu_2^L(\infty) = RT \ln\left(\frac{C_2^L(r)}{C_2^L(\infty)}\right) \qquad (6.82)$$

Similarly, if the activity coefficients in the β phase are insensitive to concentration,

$$\mu_2^\beta(r) - \mu_2^\beta(\infty) = RT \ln\left(\frac{C_2^\beta(r)}{C_2^\beta(\infty)}\right) \qquad (6.83)$$

Insertion of Equations (6.82) and (6.83) into Equation (6.79) leads to

$$RT \ln\left(\frac{C_2^\beta(r)}{C_2^\beta(\infty)}\right) - RT \ln\left(\frac{C_2^L(r)}{C_2^L(\infty)}\right) = \frac{2(\sigma - \gamma)V_S^\circ}{r} \qquad (6.84)$$

Equation (6.84) may be arranged to give the phase boundaries of the solidus and liquidus curves as

$$\frac{C_2^\beta(r)}{C_2^L(r)} = \frac{C_2^\beta(\infty)}{C_2^L(\infty)} \exp\left[\frac{2(\sigma - \gamma)V_S^\circ}{rRT}\right] \qquad (6.85)$$

Equation (6.85) shows that in this case the partition ratio is affected by both γ and σ as the system becomes finely divided.

6.5.2.3 Behavior of nanoparticles

The discussion above was directed at the case where fine particles of a β phase were in equilibrium with a bulk phase, L or α. Jesser et al. [9] have studied the behavior when both phases are small in extent. This was done by observing the melting behavior of nanoparticles of Pb–Bi alloys in a hot-stage transmission electron microscope. In this case, as melting proceeded the particles consisted of solid inside a shell of liquid until melting was complete. The Bi-rich side of the binary phase diagram is analogous to the hypothetical diagram in Figure 6.15. It was observed that the melting point of Bi was lowered for the nanoparticles, as described above. However, additional features include a dramatic

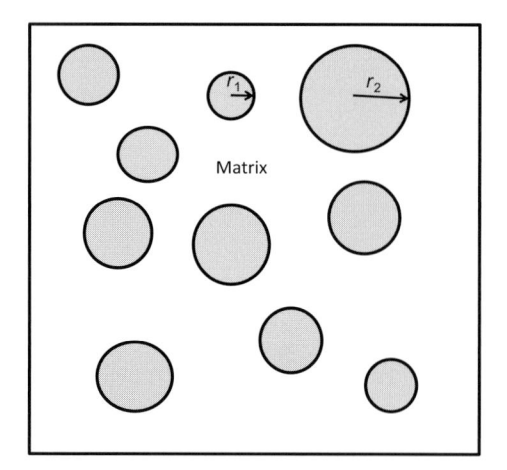

Figure 6.18 A schematic diagram of a distribution of particles of different sizes in a common matrix.

increase in solubility of Pb in solid Bi, from being negligible to as high as 20 at%. Also, the spacing between the solidus and liquidus curves decreased with decreasing particle size and collapsed down to essentially a common line for the smallest particles of radius ≈ 5 nm. Clearly, this complex behavior cannot be analyzed by simple shifts in the curves on the G versus X diagram.

6.5.3 Precipitate coarsening

There are many applications that rely on a fine dispersion of particles in a matrix to provide specific properties (mechanical, magnetic, etc.). The stability of the dispersion is important. Consider the distribution of particles in Figure 6.18. It is clear from Equation (6.75) that the activity of component 2 and its chemical potential in a small particle, e.g. of radius r_1, will be greater than that in a large particle, e.g. of radius r_2. Thus there is a chemical-potential gradient from particle 1 to particle 2. Also, there will be a higher concentration of component 2 in the matrix

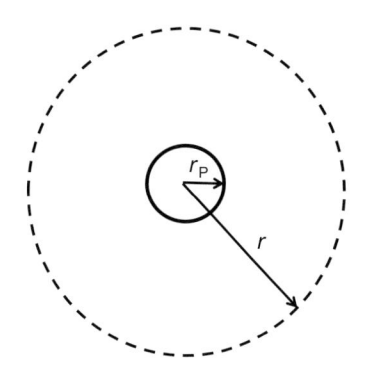

Figure 6.19 A depiction of the diffusion field around a growing or dissolving particle.

in equilibrium with particle 1 than there is in equilibrium with particle 2. This can be seen directly from Equation (6.78):

$$C_2(r_1) - C_2(r_2) = \frac{2\gamma V_2^\circ C_2(\infty)}{rRT}\left(\frac{1}{r_1} - \frac{1}{r_2}\right) \quad (6.86)$$

This concentration difference will result in diffusion of component 2 from the small particle to the larger particle. A simple diffusion analysis of the coarsening kinetics was presented by Greenwood [10]. The rate of growth or dissolution of a particle of radius r_P will be given by

$$4\pi r_P^2 \frac{dr_P}{dt} = 4\pi r^2 D_2 \frac{dC_2}{dr} \quad (6.87)$$

where dC_2/dr is the concentration gradient of solute in the surrounding matrix, Figure 6.19. If the solubility at large values of r is taken as the solubility of particles of the mean size $\bar{r}_P$, Equation (6.87) may be approximated as

$$4\pi r_P^2 \frac{dr_P}{dt} = 4\pi r^2 D_2 \frac{\Delta C_2}{r} = -4\pi r D_2[C_2(r) - C_2(\bar{r})] \quad (6.88)$$

Evaluation of the right-hand side of Equation (6.88) at $r = r_P$ yields

$$\frac{dr_P}{dt} = -\frac{D_2}{r_P}[C_2(r_P) - C_2(\bar{r}_P)] \tag{6.89}$$

Insertion of Equation (6.78) for the concentrations into Equation (6.89) yields

$$\frac{dr_P}{dt} = -\frac{2D_2 \gamma V_2^\circ C_2(\infty)}{r_P RT}\left[\frac{1}{r_P} - \frac{1}{\bar{r}_P}\right] \tag{6.90}$$

Equation (6.90) indicates two important points.

(a) Particles with radii larger than the mean have a positive growth rate (they are growing) and those with radii smaller than the mean have a negative growth rate (they are shrinking).
(b) There is a maximum growth rate, which occurs for particles with a radius exactly twice the mean. This may be seen by differentiating dr_P/dt with respect to r_P and equating it to zero.

Insertion of $\bar{r}_P = \frac{1}{2}r_P$ into Equation (6.90) yields the growth rate of the fastest growing particles as

$$\frac{dr_P}{dt} = \frac{2D_2 \gamma V_2^\circ C_2(\infty)}{r_P^2 RT} \tag{6.91}$$

This equation may be separated and integrated to give the time dependence of the most rapidly growing particles:

$$\int_{r_0}^{r_P} r_P^2 \, dr_P = \int_0^t \frac{2D_2 \gamma V_2^\circ C_2(\infty)}{RT} \, dt \tag{6.92}$$

Upon integration Equation (6.92) becomes

$$\tfrac{1}{3}r_P^3 - \tfrac{1}{3}r_0^3 = \frac{2D_2 \gamma V_2^\circ C_2(\infty)}{RT} t \tag{6.93}$$

Alternatively,

$$\tfrac{8}{3}\bar{r}_P^3 - \tfrac{8}{3}\bar{r}_0^3 = \frac{2D_2 \gamma V_2^\circ C_2(\infty)}{RT} t \tag{6.94}$$

or

$$\bar{r}_P^3 - \bar{r}_0^3 = \frac{3D_2\gamma V_2^\circ C_2(\infty)}{4RT} t \qquad (6.95)$$

The simplified analysis by Greenwood has been refined over the years. These developments have been summarized by Voorhees [11].

6.6 Summary

In this chapter the effects of surface/interface curvature on phase equilibria have been described.

The excess gas pressure inside a bubble is determined by the surface stress and is given by the Laplace equation,

$$\Delta P = \sigma \left(\frac{1}{r_1} + \frac{1}{r_2} \right) \qquad (6.96)$$

which, for the special case of a spherical bubble, is

$$\Delta P = \frac{2\sigma}{r} \qquad (6.97)$$

In the special case for which all the phases are fluids, e.g. a gas bubble in a liquid, $\sigma = \gamma$ and

$$\Delta P = \frac{2\gamma}{r} \qquad (6.98)$$

In most cases of multiphase equilibrium the contribution of σ cancels out and the surface/interface effects are determined by γ.

The increase in vapor pressure of a liquid droplet or an isotropic solid particle as the radius is decreased is given by the Kelvin equation,

$$\ln \left[\frac{P(r)}{P(\infty)} \right] = \frac{V^L}{RT} \frac{2\gamma}{r} \qquad (6.99)$$

The description of the vapor pressure of a small crystal becomes a little more complex, but the appropriate surface function to describe the equilibrium is still γ.

The lowering of the melting point of an isotropic solid as the particle radius is decreased is given by

$$T_m(\infty) - T_m(r) = \frac{2\gamma_{SL} T_m(\infty)}{r \Delta H_m} \qquad (6.100)$$

In multiphase equilibria in binary systems involving solid and liquid phases the solubility of small particles is increased. The effects of this increase have been described for the shifts in phase boundaries on binary T–X diagrams and for the phenomenon of particle coarsening.

6.7 References

[1] A. W. Adamson and A. P. Gast, *Physical Chemistry of Surfaces*, 6th edn. (New York, John Wiley & Sons, 1997), Chapter 1.
[2] L. E. Murr, *Interfacial Phenomena in Metals and Alloys* (Reading, MA: Addison-Wesley Publishing Co., 1975), Chapter 3.
[3] N. Eustathopoulos, M. G. Nicholas and B. Drevet, *Wettability at High Temperatures* (Kidlington: Elsevier Science, 1999), Chapter 3.
[4] C. A. Johnson, Generalization of the Gibbs–Thomson equation, *Surface Science*, **3** (1965), 429–444.
[5] D. A. Porter and K. E. Easterling, *Phase Transformations in Metals and Alloys* (London: Chapman & Hall, 1992), Chapter 3.
[6] J. W. Gibbs, *Collected Works*, Vol. 1 (New Haven, Yale University Press, 1957).
[7] J. W. Cahn, Surface stress and the chemical equilibrium of small crystals – 1. The case of the isotropic surface, *Acta Metall.*, **28** (1980), 1333–1338.
[8] M. Hillert and J. Agren, Effect of surface free energy and surface stress on phase equilibria, *Acta Mater.*, **50** (2002), 2429–2441.
[9] W. A. Jesser, R. Z. Shneck and W. W. Gile, Solid–liquid equilibria in nanoparticles of Pb–Bi alloys, *Phys. Rev. B*, **69** (2004), 144121-1–144121-13.

[10] G. W. Greenwood, The growth of dispersed precipitates in solutions, *Acta Metall.*, **4** (1956), 243–248.
[11] P. W. Voorhees, The theory of Ostwald ripening, *J. Statist. Phys.*, **38** (1985), 231–252.

6.8 Study problems

1. The solubility of diatomic gases, such as hydrogen, in metals is found to follow Sievert's law

$$\text{wt\% H} = k_S p_{H_2}^{1/2}$$

   Consider that a bath of liquid Fe is equilibrated with a hydrogen atmosphere at an elevated temperature and dissolves 2×10^{-2} wt% H. The liquid is rapidly cooled to 1,540 °C, at which temperature the Sievert's-law constant is 2.3×10^{-3} wt% H/atm$^{1/2}$, and hydrogen bubbles are nucleated. Calculate the minimum bubble size which is stable at a depth of 1 m below the bath surface.

   The surface energy for the H_2/Fe interface may be approximated as the surface energy of liquid iron (Table 2.1). Remember that the total pressure of the liquid at any depth is given by

$$P = P_{atm} + \rho g h$$

   where ρ is the liquid density (7.23 g/cm^3) and h is the distance below the surface. Use $\rho_{Fe(l)} = 7{,}230$ kg/m^3 and $g = 9.81$ m/s^2.

2. Consider a bath of liquid copper held at 1,400 K.
   (a) Estimate the vapor pressure of liquid copper over the bath using the entropy of vaporization and the boiling point.
   (b) Compare this value of the vapor pressure with that from the experimentally determined vapor-pressure equation

(p in atm)

$$\ln p_{Cu} = -\frac{40{,}350}{T} - 1.21 \ln T + 21.67$$

(c) Compute the equilibrium vapor pressure inside a bubble of argon gas of diameter 50 nm suspended in the liquid copper at 1,400 K. Take the surface energy of Cu(l) from Table 2.1.

(d) Compute the vapor pressure of a droplet of Cu of diameter 50 nm at the same temperature.

Use $T_b^{Cu} = 2{,}830$ K and $\Delta S_{Cu}^{Vap} = 108$ J/mole K. $V_{Cu} = 8.1$ cm^3/mole.

3. The phase diagram for the system A–B is a simple eutectic system. Below the eutectic temperature of 900 K it consists of a dilute terminal solid solution α in equilibrium with β, which is essentially pure component B with no intermediate phases. At 700 K the solubility limit of β in α is $X_B^\alpha = 0.025$. The molar volume of β is 9.5×10^{-6} m^3/mole and $\gamma_{\alpha/\beta} = 500$ mJ/m^2. Calculate the solubility limit of 50-nm-diameter particles of β in α.

4. It has been observed that very pure gold can be supercooled well below its normal melting point without freezing. Consider a bath of Au(l) that has been supercooled to 1,200 K. Calculate the critical radius for the nucleation of solid in the liquid.

Use $T_m^{Au} = 1{,}338$ K, $\Delta H_{Au}^m = 12{,}360$ J/mole and $\gamma_{SL} = 132$ mJ/m^2. The molar volume of solid Au may be taken as 10.21 cm^3/mole.

5. Calculate the air pressure inside a soap bubble with a radius of 500 μm. Assume that the surface energy of the soap solution is approximately half that of pure water.

6. Consider the shifts that occur for small solid particles on the P–T diagram for a unary system, Figure 6.12. Sketch for this system G

versus T diagrams similar to Figure 1.5 that are consistent with these observations.

7. Consider a phase diagram for a binary system, which contains a congruently melting compound AB that forms eutectic equilibria with components A and B.
    (a) Sketch the bulk phase diagram using coordinates T versus X_B.
    (b) Sketch the shifts you would expect in the diagram if the AB were present as very small particles.

7 Adsorption

The change in composition which occurs at a surface or interface is an important aspect of multicomponent systems. Enrichment of a component is termed *adsorption* (this is sometimes also called *segregation*) and denudation is termed *desorption*. This chapter describes the thermodynamics of adsorption/desorption and provides several examples of how they influence the properties of material systems.

In Chapter 2 two approaches for characterizing a surface/interface were described. In the first approach one considers the region of structural variation as a *surface phase* and treats its thermodynamic properties in the same manner as one does those of any bulk phase. The second approach, which was introduced by Gibbs [1], involves the assumption that all property changes occur at a single plane, the *Gibbs dividing surface*. All of the properties of the two phases are presumed to have their bulk values right up to the dividing surface. This approach is the easiest to use for describing the phenomena of adsorption and desorption and will be used in this chapter. Thus any extensive property may be described by

$$Q'_{tot} = Q'_\alpha + Q'_\beta + Q^S \qquad (7.1)$$

where the subscripts α and β refer to the coexisting phases and S refers to the interface. If Q' is allowed to represent numbers of moles of component A then Equation (7.1) becomes

$$n_A^{tot} = n_A^\alpha + n_A^\beta + n_A^S \qquad (7.2)$$

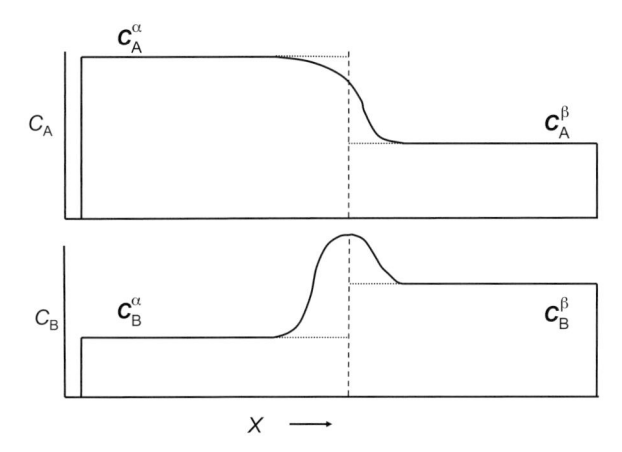

Figure 7.1 The concentration profiles of the major component A and minor component B as a function of distance for a hypothetical system A–B which has two phases α and β in contact. C_A and C_B are the concentrations of A and B, respectively, in units of moles/m^3.

This approach is illustrated for a hypothetical system A–B in Figure 7.1, which has two phases α and β in contact. C_A and C_B are the concentrations of A and B, respectively, in units of moles/m^3. Figure 7.1(a) shows the concentration profile of the major component A as a function of distance. Equation (7.2) may be used to express the number of moles at the surface as

$$n_A^S = n_A^{tot} - C_A^\alpha V^\alpha - C_A^\beta V^\beta \qquad (7.3)$$

The position of the dividing surface is generally chosen such that $n_A^S = 0$. This choice of position generally means that $n_B^S \neq 0$. This is the case in Figure 7.1(b), where the actual amount of B exceeds that represented by the area under the profile defined by the dividing surface. This means that there is an excess of B at the interface, i.e. it is *adsorbed* ($n_B^S > 0$). It is clear that, for different concentration profiles, n_B^S could be positive, negative or zero.

In describing the properties of surfaces it is generally useful to normalize them to unit area of surface. Thus the surface concentrations of A and B are defined as

$$\Gamma_A \equiv \frac{n_A^S}{A} \qquad (7.4)$$

and

$$\Gamma_B \equiv \frac{n_B^S}{A} \qquad (7.5)$$

where A is the area of the interface.

Whether a given solute will be adsorbed onto or desorbed from a given surface/interface depends on the solution thermodynamics involving the solute. The simplest approach to this problem is that described in the next section.

7.1 The Gibbs adsorption equation

The combined expression for the first and second laws for a system that includes a planar surface/interface is

$$dE' = T\,dS' - P\,dV' + \sum_i \mu_i\,dn_i + \gamma\,dA \qquad (7.6)$$

where

$$E' = E'_\alpha + E'_\beta + E^S \qquad (7.7)$$

$$S' = S'_\alpha + S'_\beta + S^S \qquad (7.8)$$

$$V' = V'_\alpha + V'_\beta \qquad (7.9)$$

and

$$n_i = n_i^\alpha + n_i^\beta + n_i^S \qquad (7.10)$$

The integrated form of the combined first and second laws for the entire system shown in Figure 7.1 is

$$E' = TS' - P_\alpha V'_\alpha - P_\beta V'_\beta + \sum_i n_i \mu_i + \gamma A \qquad (7.11)$$

where E' and S' are the extensive internal energy and entropy of the system, respectively. The chemical potential μ_i is taken as constant throughout the system. The internal energies of the α and β phases are given by

$$E'_\alpha = TS'_\alpha - P_\alpha V'_\alpha + \sum_i n_i^\alpha \mu_i \tag{7.12}$$

and

$$E'_\beta = TS'_\beta - P_\beta V'_\beta + \sum_i n_i^\beta \mu_i \tag{7.13}$$

The internal energy associated with the interface is the difference between the internal energy of the system and that of the α and β phases, as determined by the position of the Gibbs dividing surface,

$$E^S = E' - E'_\alpha - E'_\beta \tag{7.14}$$

which yields

$$E^S = TS^S + \sum_i n_i^S \mu_i + \gamma A \tag{7.15}$$

Differentiation of Equation (7.15) for an infinitesimal change in state of the system yields

$$dE^S = T\,dS^S + S^S\,dT + \sum_i n_i^S\,d\mu_i + \sum_i \mu_i\,dn_i^S + \gamma\,dA + A\,d\gamma \tag{7.16}$$

This may be compared with the combined expression for the first and second laws applied to the surface, which is

$$dE^S = T\,dS^S + \sum_i \mu_i\,dn_i^S + \gamma\,dA \tag{7.17}$$

which indicates that

$$S^S\,dT + \sum_i n_i^S\,d\mu_i + A\,d\gamma = 0 \tag{7.18}$$

Dividing through Equation (7.18) by the interfacial area, A, and rearranging yields

$$d\gamma = -s^S dT - \sum_i \Gamma_i d\mu_i \qquad (7.19)$$

which is the *Gibbs adsorption isotherm*. In most applications the temperature is held constant so that Equation (7.19) is reduced to

$$d\gamma = -\sum_i \Gamma_i d\mu_i \qquad (7.20)$$

Equation (7.20) indicates that the surface concentrations will be determined by the effects of chemical potentials on the surface energy. This will be illustrated by several examples.

7.1.1 Applications of the Gibbs adsorption equation

7.1.1.1 Binary solutions at constant temperature

For a binary solution at constant temperature, Equation (7.20) may be expanded as

$$d\gamma = -\Gamma_1 d\mu_A - \Gamma_2 d\mu_B \qquad (7.21)$$

In most instances the position of the Gibbs dividing surface may be chosen such that $\Gamma_A = 0$ and Equation (7.21) may be rearranged to express the surface excess of the solute as

$$\Gamma_B = -\left(\frac{\partial \gamma}{\partial \mu_B}\right)_T \qquad (7.22)$$

Thus, if increasing the chemical potential (i.e. the concentration) of component B decreases the interfacial energy that species will be *adsorbed*, whereas if doing this increases γ it will be *desorbed*.

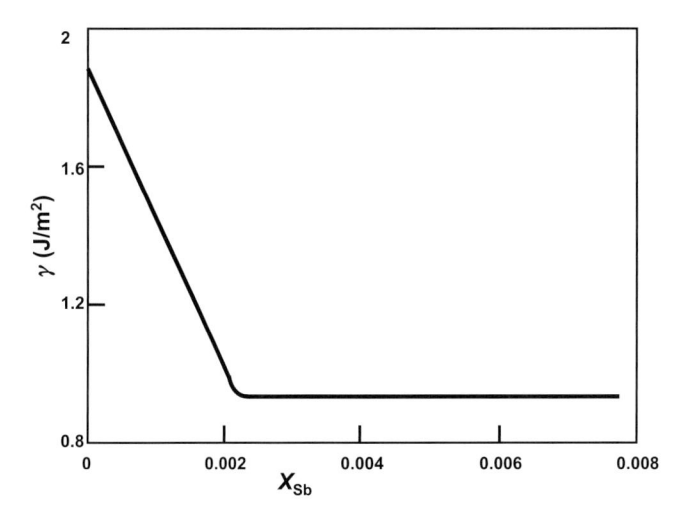

Figure 7.2 The effect of the mole fraction of antimony (X_{Sb}) on the surface energy of dilute Cu–Sb solid solutions at 950 °C. (After Reference [2] – individual data points are not shown.)

The chemical potential of the solute may be expressed in terms of the activity as

$$\mu_B = \mu_B^\circ + RT \ln a_B \tag{7.23}$$

and

$$\Gamma_B = -\frac{1}{RT}\left(\frac{\partial \gamma}{\partial \ln a_B}\right)_T \tag{7.24}$$

If the solution is dilute enough that the solute obeys Henry's law, Equation (1.54), then

$$\Gamma_B = -\frac{1}{RT}\left(\frac{\partial \gamma}{\partial \ln X_B}\right)_T = -\frac{X_B}{RT}\left(\frac{\partial \gamma}{\partial X_B}\right)_T \tag{7.25}$$

Figure 7.2 shows the dependence of the surface energy of the Cu–Sb system at 950 °C as a function of the concentration of antimony in the alloys [2]. The rapid decrease in γ with small Sb additions is the result of Sb being a surface-active element. The horizontal section of the plot

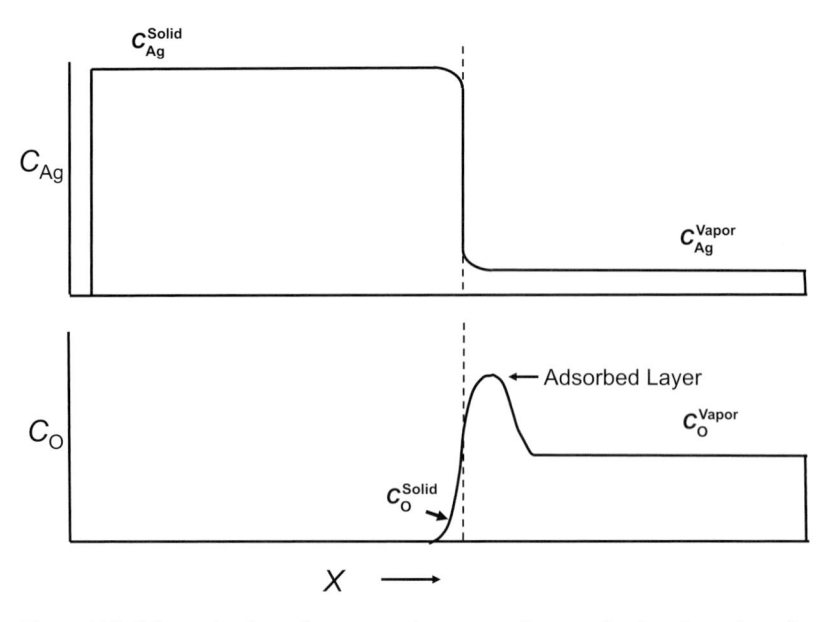

Figure 7.3 Schematic plots of concentration versus distance for the adsorption of oxygen on silver.

at higher Sb concentrations is the result of saturation of the surface with Sb. In general, saturation is presumed to occur when one atomic layer (monolayer) of solute occupies the surface.

7.1.1.2 Adsorption from the gas phase at constant temperature

If a species, e.g. oxygen, is adsorbing from the gas phase Equation (7.23) takes the form

$$\mu_{O_2} = \mu^{\circ}_{O_2} + RT \ln p_{O_2} \qquad (7.26)$$

and the Gibbs adsorption isotherm, Equation (7.22) becomes

$$\Gamma_{O_2} = -\frac{1}{RT}\left(\frac{\partial \gamma}{\partial \ln p_{O_2}}\right)_T \qquad (7.27)$$

Figure 7.3 shows schematic representations of the concentration profiles when oxygen adsorbs on silver. The position of the dividing surface

7.2 THE LANGMUIR ADSORPTION EQUATION

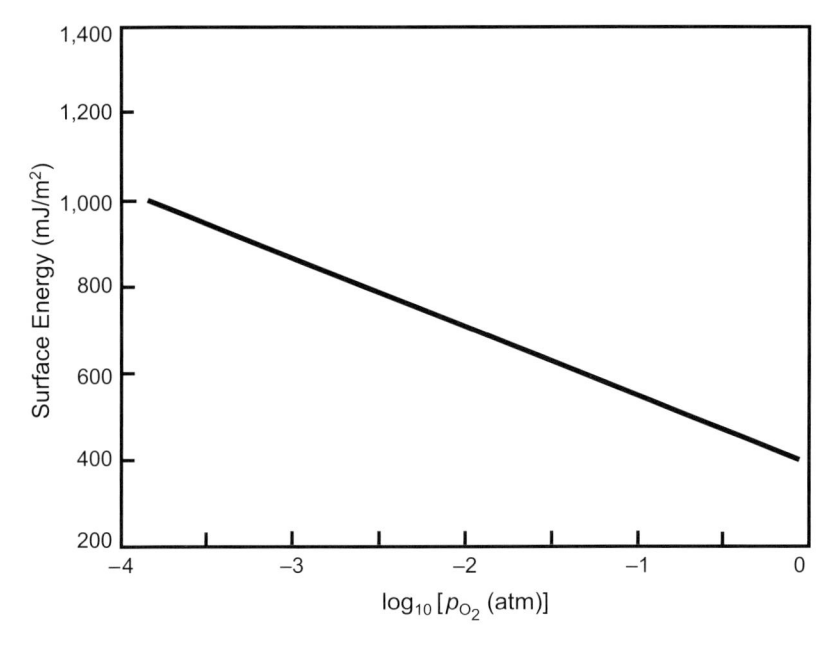

Figure 7.4 A plot showing the effect of oxygen partial pressure on the surface energy of silver. (After Reference [2] – individual data points are not shown.)

was chosen so that $\Gamma_{Ag} = 0$. Figure 7.4 shows how the surface energy of silver varies with the logarithm of the partial pressure of oxygen in the ambient atmosphere. The linear form of this plot indicates that the oxygen excess is constant over the range of pressures studied.

Figure 7.5 shows a similar plot for oxygen adsorbing on copper [3]. The discontinuity in this plot at a p_{O_2} below the bulk dissociation pressure is the result of the surface energy stabilization of Cu_2O described in Section 5.3.1.1.

7.2 The Langmuir adsorption equation

The saturation of a surface by an adsorbed species, e.g. as in Figure 7.2, is not described by the Gibbs adsorption isotherm. Several isotherms

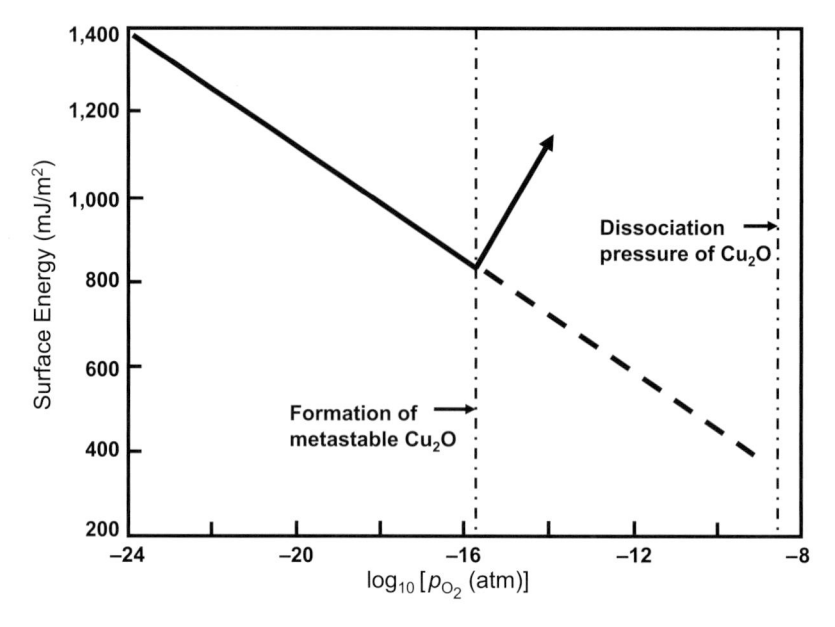

Figure 7.5 A plot showing the effect of oxygen partial pressure on the surface energy of copper. (After Reference [3] – individual data points not shown.)

have been used to describe this behavior. The simplest of these is the Langmuir adsorption isotherm [4]. The isotherm, which was originally used to describe the adsorption of gases, may be derived using either reaction kinetics or statistical mechanics [5]. The former approach will be followed here.

Consider the adsorption of water vapor on a solid surface. The adsorption may be written in the form of a reaction

$$H_2O(g) + V^S = H_2O^S \qquad (7.28)$$

where V^S represents a vacant surface site and H_2O^S represents an adsorbed water molecule. It is assumed that there is a fixed number of sites available for adsorption and that the adsorbed species do not interact. At any stage of the reaction the concentration of V^S (expressed per unit area) will be

$$C_{VS} = (\Gamma_{H_2O}^{sat} - \Gamma_{H_2O}) \tag{7.29}$$

where $\Gamma_{H_2O}^{sat}$ is the surface excess at saturation and, therefore, is the total number of adsorption sites per unit area. The rate of the forward reaction may be expressed as

$$\text{Rate}_f = k_f p_{H_2O}(\Gamma_{H_2O}^{sat} - \Gamma_{H_2O}) \tag{7.30}$$

and the rate of the backward reaction will be

$$\text{Rate}_b = k_b \Gamma_{H_2O} \tag{7.31}$$

where k_f and k_b are the rate constants for the forward and backward reactions, respectively. At equilibrium the two rates will be equal:

$$k_b \Gamma_{H_2O} = k_f p_{H_2O}(\Gamma_{H_2O}^{sat} - \Gamma_{H_2O}) \tag{7.32}$$

Rearrangement of Equation (7.32) yields

$$\frac{\Gamma_H}{p_{H_2O}(\Gamma_{H_2O}^{sat} - \Gamma_{H_2O})} = \frac{k_f}{k_b} = K \tag{7.33}$$

where K is the equilibrium constant for the adsorption reaction, Equation (7.28). The fraction of sites covered by water molecules is given by

$$\Theta = \frac{\Gamma_{H_2O}}{\Gamma_{H_2O}^{sat}} \tag{7.34}$$

The fractional coverage Θ may be expressed as a function of partial pressure from Equation (7.33) as

$$\frac{1}{K p_{H_2O}} = \frac{\Gamma_{H_2O}^{sat}}{\Gamma_{H_2O}} - 1 = \frac{1}{\Theta} - 1 \tag{7.35}$$

or

$$\Theta = \frac{K p_{H_2O}}{1 + K p_{H_2O}} \tag{7.36}$$

which is the *Langmuir adsorption isotherm*.

The Langmuir equation has also been used to describe the adsorption of solid elements from solution onto free surfaces, e.g. Figure 7.2, and to grain boundaries. For this case the isotherm may be derived in a similar fashion to Equation (7.36) by considering the adsorption of a component from solution onto a free surface or grain boundary. Consider the case of sulfur adsorption:

$$\underline{S} + V^S = S^S \tag{7.37}$$

where $\underline{S}$ represents a sulfur atom in solid solution, V^S represents a vacant surface site and S^S represents a sulfur atom adsorbed on the surface.

The equilibrium constant (this case is analogous to Equation (7.33)) will be

$$K = \frac{a_{S^S}}{a_{\underline{S}} a_{V^S}} \tag{7.38}$$

Replacing the activities of S^S and V^S by their surface excesses and defining a Henrian standard state for $\underline{S}$ yields

$$K = \frac{\Gamma_S}{X_S(\Gamma_S^{sat} - \Gamma_S)} \tag{7.39}$$

which may be rearranged to give the fractional coverage

$$\Theta = \frac{K X_S}{1 + K X_S} \tag{7.40}$$

It should be noted that Equations (7.36) and (7.40) may be arranged, respectively, to

$$\frac{\Theta}{1-\Theta} = p_{H_2O} K \qquad (7.41)$$

and

$$\frac{\Theta}{1-\Theta} = X_S K \qquad (7.42)$$

The temperature dependence of K may be expressed using the van 't Hoff isotherm, Equation (1.87), as

$$K = \exp\left(-\frac{\Delta G^\circ_{ads}}{RT}\right) = \exp\left(\frac{\Delta S^\circ_{ads}}{R}\right) \exp\left(-\frac{\Delta H^\circ_{ads}}{RT}\right) \qquad (7.43)$$

Thus a plot of $\ln[\Theta/(1-\Theta)]$ versus reciprocal temperature at constant p or X will yield a slope of $-\Delta H^\circ_{ads}/R$, where ΔH°_{ads} is the *heat of adsorption*. Grabke [6] has shown that Equation (7.42) describes the adsorption of a number of solutes to the surfaces and grain boundaries of iron and has used Equation (7.43) to obtain the corresponding heats of adsorption.

7.3 The effects of adsorption on the fracture of solids

There are numerous examples of fracture being affected by adsorption phenomena. In this section two important cases will be described. These are the effect of water vapor on the fracture of ceramics and the effect of grain-boundary segregation on the fracture of metals.

7.3.1 The effect of water vapor on the fracture of ceramics

Static fatigue of ceramics occurs by the growth of subcritical surface cracks until they satisfy the Griffith criterion for fracture, i.e. the cracks reach a critical size [7]. This process can be strongly influenced by the

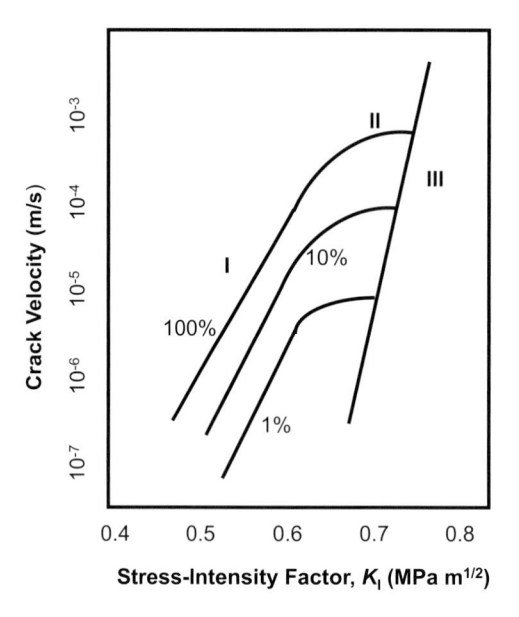

Figure 7.6 Effects of stress and gaseous moisture content on the crack-growth velocity in soda-lime glass.

adsorption of water at the crack tip. The effect of water vapor on slow crack propagation in glass was reported in a classic study by Wiederhorn [8]. The velocity of crack propagation in soda-lime-silica glass microscope slides was measured using a double-cantilever cleavage technique. Selected results from this study are plotted in Figure 7.6, which shows the crack velocity as a function of stress-intensity factor for several different levels of relative humidity in a nitrogen carrier gas.

In region I the crack velocity depends exponentially on the stress-intensity factor and is also dependent on the partial pressure of H_2O. In region II the crack velocity is nearly independent of K_I but increases as the partial pressure of water increases. In region III the crack velocity depends even more strongly on K_I and is independent of the partial pressure of water vapor. The cracking behavior of Al_2O_3 (sapphire) was

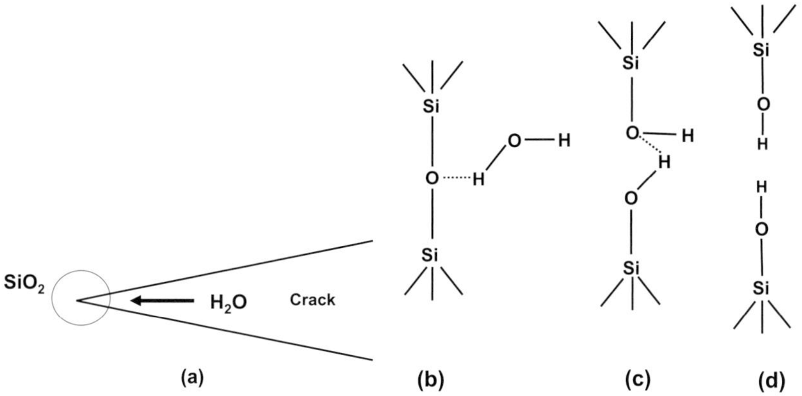

7.3 EFFECTS OF ADSORPTION ON FRACTURE OF SOLIDS 197

Figure 7.7 A schematic diagram of water diffusing to a crack tip (a) and the sequence of interactions of the water with the bonds in the silica at the tip, (b)–(d).

found to be qualitatively similar to that for glass [9]. Therefore, vitreous and crystalline ceramic behave similarly.

The behavior in regions I and II was explained using the earlier stress corrosion theory of Charles and Hillig [10]. The environmental dependence of the crack velocity arises from the dependence of the activation on the chemical potential of the water vapor in the gas.

A mechanistic description of the behavior in region I for vitreous silica was provided by Michalske and Freiman [11]. This mechanism is shown schematically in Figure 7.7. When a tensile stress is applied to vitreous silica the Si—O bonds are stretched. This is particularly important in the region of the crack tip (the circled region of Figure 7.7(a)), where highly concentrated strain fields are produced. When water is adsorbed at the crack tip according to Equation (7.28), it can attach to a bridging Si—O bond, as indicated in Figure 7.7(b). A reaction involving the bonding of a hydrogen ion (proton) to an oxide ion from the Si—O bond and transfer of an electron from the oxygen in the water molecule to the Si ion then occurs. The net effect is that two new bonds have replaced the original Si—O bond (Figure 7.7(c)). Finally, the hydrogen bond between the oxygen (originally from the H_2O) and the transferred hydrogen breaks under the applied stress and the crack has propagated one atomic distance, Figure 7.7(d). This model explains the increase in crack velocity with increased stress intensity (greater bond stretching) and partial pressure of water vapor (increased surface coverage, e.g. Equation (7.36)).

In region II the mechanism of fracture changes from the reaction-rate-limited process of region I to a transport-rate-limited process. The reason is that the stress-activated process at the crack tip has become faster than the rate at which water vapor can diffuse to the crack tip.

In region III, the crack velocity is independent of the partial pressure of water. The shape of the crack-growth curve in region III has been explained by Wiederhorn *et al.* [12] in terms of electrostatic interactions between the environment and charges that develop at the crack tip during fracture.

7.3.2 The effect of grain-boundary segregation on the fracture of metals

The adsorption of a solute at the grain boundaries in a metal or alloy (*grain-boundary segregation*) can have dramatic effects on mechanical behavior. In some cases a stressed polycrystalline metal (alloy) undergoes substantial plastic deformation before fracture, as indicated by the schematic stress–strain curve in Figure 7.8(a). However, the same material can fail in a totally brittle fashion, with little or no plastic deformation, if a solute has segregated to the grain boundaries as indicated in Figure 7.8(b). Figures 7.9–7.11 show fracture surfaces corresponding to the stress–strain curves in Figure 7.8. Figure 7.9 shows the fracture surface of a steel that has undergone ductile fracture corresponding to Figure 7.8(a). Figure 7.10 shows the fracture surface of a steel that has undergone brittle cleavage fracture corresponding to Figure 7.8(b).

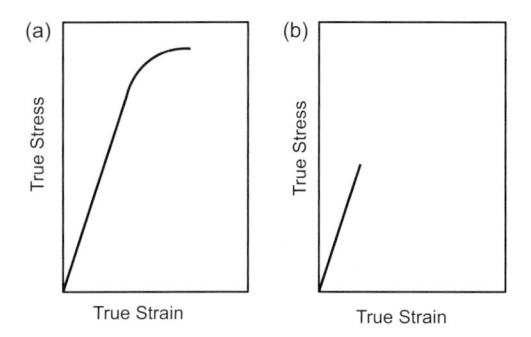

7.3 EFFECTS OF ADSORPTION ON FRACTURE OF SOLIDS 199

Figure 7.8 Schematic stress–strain curves for a ductile metal (a) and a metal with embrittled grain boundaries (b).

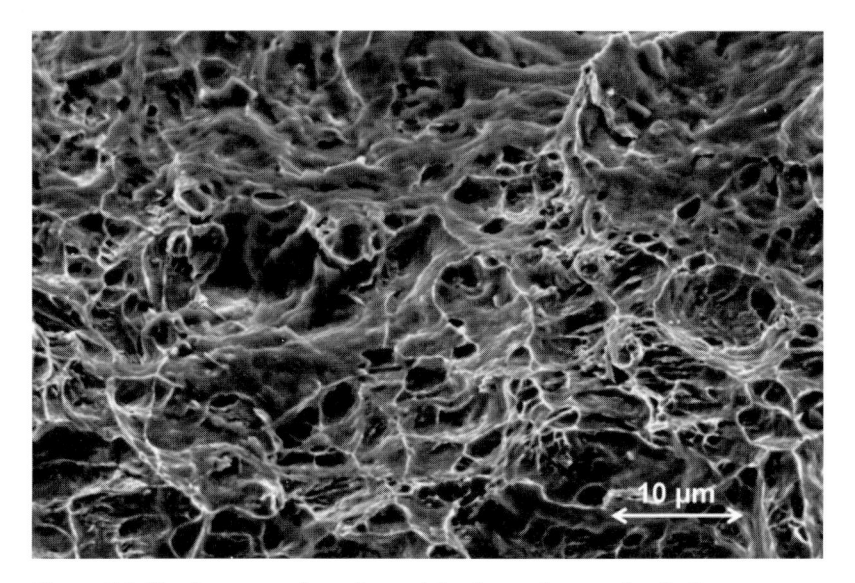

Figure 7.9 The fracture surface of a steel that has undergone ductile fracture corresponding to Figure 7.8(a). (Photograph courtesy of Dr. M. Hua.)

Figure 7.11 shows the fracture surface of an alloy that is similar to the ductile alloy in Figure 7.9 but has a solute adsorbed at its grain boundaries and has undergone brittle intergranular fracture with a stress–strain curve also corresponding to Figure 7.8(b). The latter behavior is the

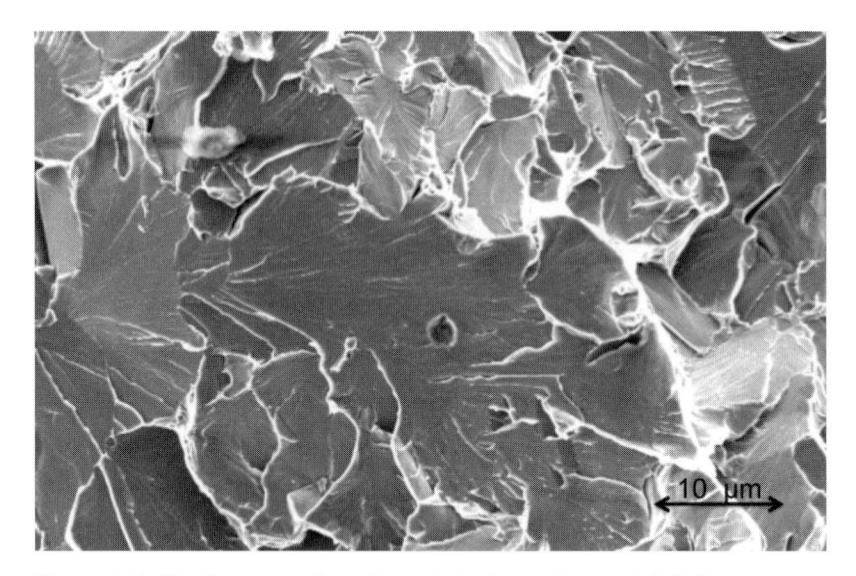

Figure 7.10 The fracture surface of a steel that has undergone brittle fracture corresponding to Figure 7.8(b). (Photograph courtesy of Dr. M. Hua.)

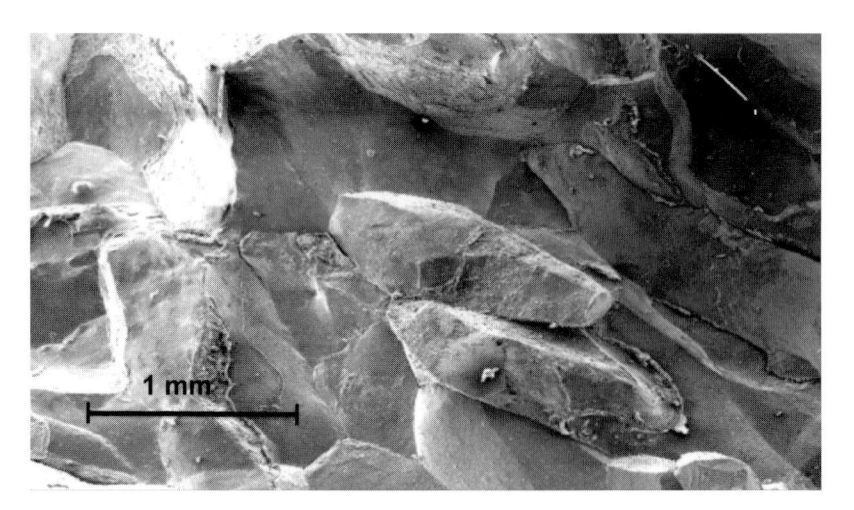

Figure 7.11 The fracture surface of an alloy that has undergone brittle intergranular fracture, also corresponding to Figure 7.8(b). (Photograph courtesy of Dr. M. Hua.)

focus of this section. The subject of grain-boundary cracking has been reviewed by Shewmon [13].

The mechanisms for the embrittling effects are not completely understood. The fundamental thermodynamic quantity is the work of separation. As a grain-boundary crack propagates grain-boundary area is destroyed and two free surfaces are created. The work associated with this process is the *ideal work of separation*,

$$w'_{Sep} = 2\gamma_{Surf} - \gamma_{gb} \tag{7.44}$$

Generally the experimental work of fracture is much larger than the ideal work of separation (except for an ideally brittle solid) because of plastic deformation of the solid prior to fracture. However, it has been found that the experimental work of fracture scales with w'_{Sep}.

Therefore, the question of how grain-boundary segregation decreases w'_{Sep} arises. Two general mechanisms have been proposed in order to explain these effects. One is based on Gibbsian adsorption (segregation) to both grain boundaries and free surfaces, with the surface energy being lowered more than the grain-boundary energy. This will be termed the *surface-energy effect*. The other mechanism involves segregation of solutes to the grain boundary because they do not fit well in the matrix, i.e. they have low solubility. In this case solutes with weak bonding weaken the grain boundary and those with strong bonding strengthen it. The second mechanism, which does not explicitly involve the grain boundary or surface energies, will be termed the *low-solubility effect*.

7.3.2.1 The surface-energy effect

According to Equation (7.25), if a dilute solute being adsorbed onto the grain boundary, i.e. Γ_B is positive, it decreases γ_{gb}, which should increase w'_{Sep}. However, the solute may also segregate to the newly created surfaces. Shewmon [13] has proposed that the fact that segregation

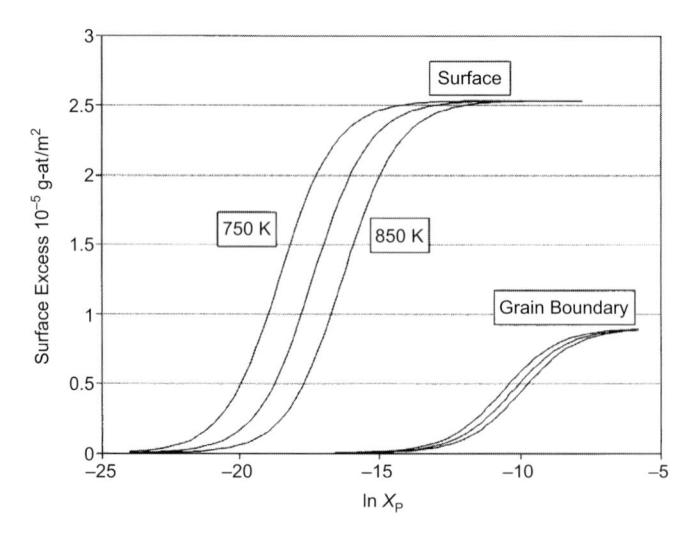

Figure 7.12 Calculated surface excesses of phosphorus on iron surfaces and grain boundaries in the range 750–850 K. (From [15], reprinted with permission of Springer.)

occurs suggests a reduction in the grain-boundary energy. However, the fact that the cracks easily follow the grain boundaries suggests that grain-boundary segregation reduces the energy of the free surface. There are not many systems for which complete data sets exist. One exception is the adsorption of phosphorus to the grain boundaries and free surfaces of Fe-base alloys. Shewmon [13] uses literature data to show that, in the temperature range in which P is known to embrittle ferrous alloys, the surface energy decreases much more strongly than the grain-boundary energy with increasing phosphorus concentration in the alloy. Figure 7.12 shows values, extrapolated from higher temperature, for the surface excess of phosphorus, Γ_P, as a function of the phosphorus content in iron. It is observed that the phosphorus is adsorbed more strongly to the free surface than to the grain boundary. Figure 7.13 shows the corresponding effect of alloy content on the surface and grain-boundary energies. It is clear that adding phosphorus dramatically decreases the surface energy

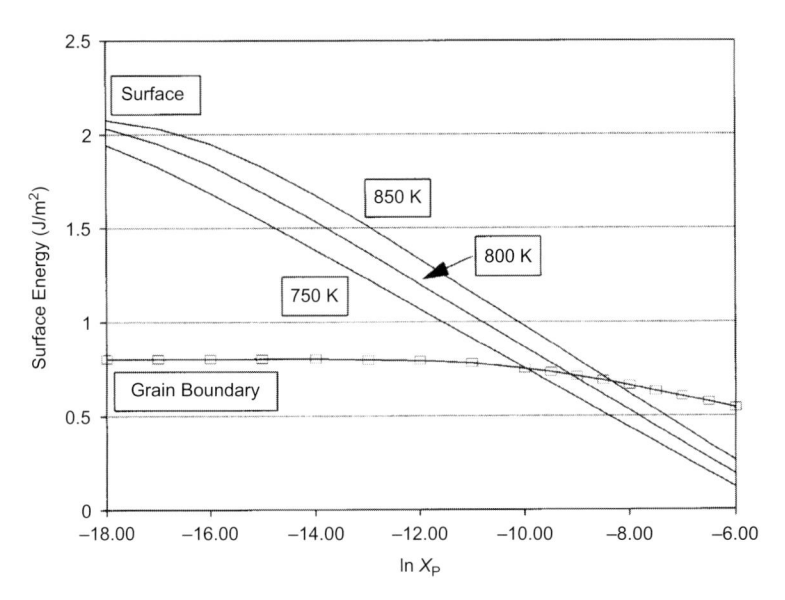

Figure 7.13 Calculated surface and grain boundary energies for Fe–P alloys corresponding to Figure 7.12. (From [15], reprinted with permission of Springer.)

but produces only a modest decrease in the grain-boundary energy. Similar results have also been reported for tin additions to iron [6, 14].

The detailed segregation mechanisms will depend on the temperature and strain rate which are used in the experiment [15, 16]. There are two limiting cases:

(1) high temperature, low strain rate, where the adsorption of solute at the free surfaces created by grain-boundary fracture is controlled by equilibrium with the bulk grains; and
(2) low temperature, high strain rate, where there is no exchange of species adsorbed on the grain boundary and those in the bulk grains.

These two cases may be understood from the schematic diagrams in Figure 7.14. Figure 7.14(a) shows a crack propagating along a grain

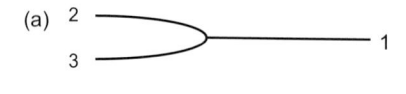

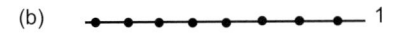

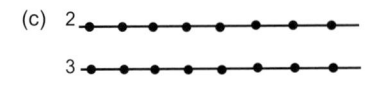

Figure 7.14 (a) A schematic diagram of a crack propagating along a grain boundary. (b) A segment of the grain boundary ahead of the crack with a solute B (black dots) adsorbed. (c) Two free surfaces created under the conditions of limiting case (1), adsorption equilibrium. (d) Two free surfaces created under the conditions of limiting case (2), where there is no time for the adsorbed solute to re-equilibrate with the matrix.

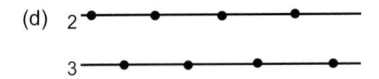

boundary. Figure 7.14(b) shows a segment of grain boundary ahead of the crack with a solute B (black dot) adsorbed. Figure 7.14(c) represents the two free surfaces created under the conditions of limiting case (1), where the adsorbed solute has been able to re-equilibrate with the matrix. For this case

$$w'_{Sep} = 2\gamma_{Surf}^{EqAd} - \gamma_{gb}^{EqAd} \qquad (7.45)$$

where the superscript EqAd refers to equilibrium adsorption.

Figure 7.14(d) represents the two free surfaces created under the conditions of limiting case (2), where there is no time for the adsorbed solute to re-equilibrate with the matrix. The only solute on the free surfaces is that remaining from the grain-boundary segregant. For this case

$$w'_{Sep} = 2\gamma_{Surf}^{MetaAd} - \gamma_{gb}^{EqAd} \qquad (7.46)$$

where the superscript MetaAd refers to metastable adsorption.

The limiting cases may be represented on the plot of surface/grain-boundary excess versus chemical potential in Figure 7.15. The starting point is the same for both processes on the isotherm for the segregated grain boundary. For case (1) the path is given by the vertical

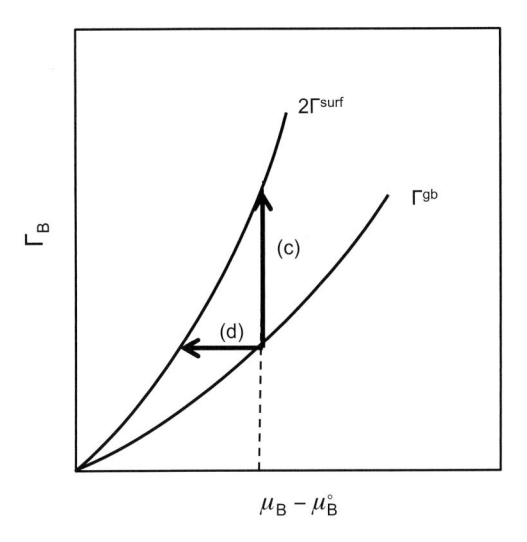

Figure 7.15 A plot of surface/grain-boundary excess versus chemical potential. The vertical arrow corresponds to process (c) in Figure 7.14. The horizontal arrow corresponds to process (d) in Figure 7.14.

arrow (c) and occurs at constant chemical potential. For case (2) the path is given by the horizontal arrow (d) and occurs for a constant number of segregated atoms, i.e. $\Gamma_{gb} = 2\Gamma^{Surf}$.

7.3.2.2 The low-solubility effect

Seah [17] has shown that there is a correlation between the tendency for a solute to segregate to a grain boundary and its solubility in the alloy matrix. The lower the solubility the greater the tendency for segregation. This can be expressed by an enrichment factor β,

$$\beta = \frac{X_B^{gb}}{X_B^{Alloy}} \propto \frac{1}{X_B^{Sol}} \qquad (7.47)$$

where X_B^{Sol} is the solubility. The effect is based mainly on the strain energy of the misfitting solute being the driving force for adsorption. Thus, in principle, a solute could segregate to a grain boundary even if it increased the grain-boundary energy, as long as it decreased the overall free energy of the system. In this situation the increased grain-boundary energy could lower w'_{Sep}, depending on the behavior of the free surface.

The weakening of the grain boundary in this case is explained in terms of bonding. Seah [15, 17] presents a plot of ΔH^{subl} versus atomic diameter for a large number of elements. Since the heat of sublimation is related to bond energies as described in Section 1.2.2,

$$\varepsilon_{\text{AA}} = -\frac{2\Delta H_A^{\text{subl}}}{N_0 Z_A} \tag{7.48}$$

and

$$\varepsilon_{\text{BB}} = -\frac{2\Delta H_B^{\text{subl}}}{N_0 Z_B} \tag{7.49}$$

From Equations (7.48) and (7.49) it is argued that if ΔH_B^{subl} is smaller than ΔH_A^{subl} segregation of B to grain boundaries in A will weaken them. This is the case for elements such as Sn, S, P and Si in iron, and these elements are known to cause grain-boundary embrittlement in iron [14]. Conversely, if ΔH_B^{subl} is larger than ΔH_A^{subl} segregation of B to grain boundaries in A will strengthen them. This is the case for elements such as C and Mo in iron, and these elements are known to increase grain-boundary cohesion in iron [14].

7.3.2.3 Case study: "green steels"

Traditionally lead has been added to steels when machinability is an important property [18]. The lead additions reduce the cutting forces and result in longer tool life. However, the extreme toxicity of lead has made the processing of these steels and the disposal of cutting waste a significant environmental problem. It has generally been accepted that lead improved machinability by forming Pb inclusions and soft MnS inclusions that were associated with the lead. However, a study by DeArdo and Garcia [18] using atom-probe field-ion microscopy (APFIM) showed that grain-boundary segregation of lead was the principal source of the improved machinability, i.e. increased brittleness, of the steels. This discovery suggested that other elements that segregate to grain boundaries might be used to replace the lead. Further APFIM investigations indicated that tin would segregate to steel grain boundaries in a similar manner to lead. Subsequent quantitative atom-probe analysis [19] showed that carbon, manganese, phosphorus and tin were segregated. The surface excesses of C, Mn and P were greater than that of Sn but the relative segregation β (grain-boundary versus matrix concentration) was larger for the Sn. Indeed, the Sn-containing steel exhibited machinability at least equivalent to that of Pb-bearing steel [18]. Furthermore, Sn is essentially non-toxic.

7.4 Summary

In this chapter adsorption onto free surfaces and grain boundaries has been described. The effects of water adsorption on slow crack growth in ceramics and of the adsorption of dilute solutes in alloys on grain-boundary embrittlement have been described. Finally, a case study of the use of adsorption phenomena in a beneficial way to produce a "green steel" for machining applications was presented.

7.5 References

[1] J. W. Gibbs, *Collected Works*, Vol. 1 (New Haven, Yale University Press, 1957).
[2] J. M. Blakely, *Introduction to the Properties of Crystal Surfaces* (Oxford, Pergamon Press, 1973).
[3] M. McLean and E. D. Hondros, Interfacial energies and chemical compound formation, *J. Mater. Sci.*, **8** (1973), 349–351.
[4] I. Langmuir, The adsorption of gases on plane surfaces of glass, mica and platinum, *J. Amer. Chem. Soc.*, **40** (1918), 1361–1368.

[5] A. W. Adamson and A. P. Gast, *Physical Chemistry of Surfaces* (New York, John Wiley & Sons, 1997), Chapter XVII.

[6] H. J. Grabke, Surface and grain boundary segregation on and in iron and steels, *Iron Steel Inst. Japan Int.*, **29** (1989), 529–538.

[7] A. A. Griffith, Phenomena of rupture and flow in solids, *Phil. Trans. Roy. Soc. (London)*, **221A** (1920), 163–198.

[8] S. M. Wiederhorn, Influence of water vapor on crack propagation in soda-lime glass, *J. Amer. Ceramic Soc.*, **50** (1967), 407–414.

[9] S. M. Wiederhorn, Moisture assisted crack growth in ceramics, *Int. J. Fracture Mech.*, **4** (1968), 171–177.

[10] R. J. Charles and W. B. Hillig, The kinetics of glass failure by stress corrosion, in *Symposium sur la résistance mécanique du verre et les moyens de l'améliorer*, Florence (Charleroi: Union Scientifique Continentale du Verre, 1962), pp. 511–527.

[11] T. A. Michalske and S. W. Freiman, A molecular mechanism for stress corrosion in vitreous silica, *J. Amer. Ceramic Soc.*, **66** (1983), 284–288.

[12] S. M. Wiederhorn, S. W. Freiman, E. R. Fuller and C. J. Simmons, Effects of water and other dielectrics on crack growth, *J. Mater. Sci.*, **17** (1982), 3460–3478.

[13] P. G. Shewmon, Grain boundary cracking, *Metall. Mater. Trans. B*, **29** (1998), 509–518.

[14] E. D. Hondros, M. P. Seah, S. Hofmann and P. Lejcek, Interfacial and surface microchemistry, in *Physical Metallurgy*, 4th edn, eds. R. W. Cahn and P. Haasen (Amsterdam: Elsevier Science, 1996), Chapter 13, pp. 1202–1289.

[15] M. P. Seah, Adsorption-induced interface decohesion, *Acta Metall.*, **28** (1980), 955–962.

[16] J. P. Hirth and J. R. Rice, On the thermodynamics of adsorption at interfaces as it influences decohesion, *Metall. Trans. A*, **11** (1980), 1501–1511.

[17] M. P. Seah, Grain boundary segregation, *J. Phys. F*, **10** (1980), 1043–1064.

[18] A. J. DeArdo and C. I. Garcia, Tin-bearing free machining steel, United States Patent No. 5,961,747, October 5, 1999.

[19] C. I. Garcia, M. J. Hua, M. K. Miller and A. J. DeArdo, Application of grain boundary engineering in lead-free "green steel", *ISIJ Int.*, **43** (2003), 2023–2027.

7.6 Study problem

Consider the effect of Sb on the surface energy of Cu in Figure 7.2. Estimate the surface excess of Sb, Γ_{Sb}, for a mole fraction of Sb in the alloy of 0.001.

8 Adhesion

Many materials are used with a layer of dissimilar material on the surface. Coatings have been used for centuries for embellishment or for protection of a substrate that is adequate in all other ways, usually providing shape, stiffness or strength. Figure 8.1(a) is a schematic representation of a coating. Coatings include paints, polymeric coatings, metallic coatings and ceramic coatings. Also, many important metallic materials develop an external surface film during service. Important examples include passive films that provide protection against aqueous corrosion (e.g. chromia films on stainless steels) [1] and oxides that form when a metal is exposed to oxygen at high temperatures [2]. Also, adhesives are often used to join two substrates, which may be of the same or dissimilar materials [3]. Figure 8.1(b) is a schematic representation of an adhesive bonding two substrates. An underlying principle, which is important for all these coatings, films and adhesives, is the adhesion of the surface layer to the substrate(s). In most instances, where the surface layer provides a useful property, strong adhesion to the substrate is desired. However, there are situations where lack of adhesion is desirable (e.g. "descaling" of steels following heat treatment).

An understanding of the fundamentals governing adhesion is important regardless of the application. These involve two opposing effects: *elastic strain energy* in the layers, which results from stresses in the layers and provides the driving force for delamination; and the layer's *adherence*, which is the resistance to separation at the interface between two layers and is intimately related to the work of adhesion (W_{ad}), which was described in Chapter 3.

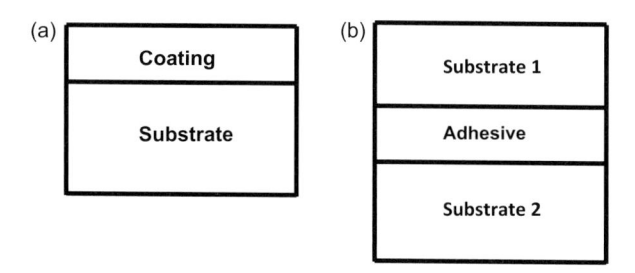

Figure 8.1 A schematic diagram of two multilayer systems, (a) a thin film on a thick substrate and (b) a thin adhesive layer bonding two substrates.

In this chapter the origins of stresses in films and the relation of the fracture energy (Γ) to the work of adhesion in bilayer and multilayer systems are described. The effects of adsorption on film adherence are described, and a case study of the relevance of film adherence to advanced superalloys is presented.

8.1 The origin of stresses in multilayer systems

The determination of the elastic strain energy requires an understanding of the stresses which may develop in the layers.

8.1.1 Formation stresses

In some cases stresses are generated in multilayer systems by the manner in which the systems are formed. For example, the volume changes which occur during the curing of an adhesive can result in stresses in the adhesive. Similarly, a coating deposited by plasma spraying may be stressed by the volume changes associated with the solidification of the deposited liquid droplets (*splats*). Also, a surface film growing by reaction (e.g. growth of an oxide) can develop stresses as a result of the location of the reaction site [2].

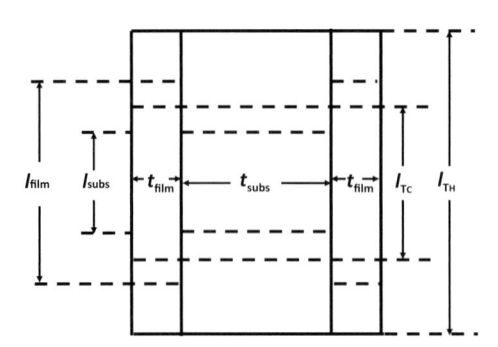

Figure 8.2 A schematic diagram showing the development of thermal stresses in a bimaterial system.

8.1.2 Thermal stresses

Even when there are no formation stresses, stresses can be generated during a temperature change because of the difference in coefficient of thermal expansion (CTE) of the two layers. A derivation for the stresses generated in a bimaterial strip subjected to a change in temperature, ΔT, was originally developed by Timoshenko [4]. The following derivation is based on Timoshenko's, for the specific case of a film on both sides of a substrate and thus no bending of the bimaterial system. The specimen is shown in one dimension in Figure 8.2. At the higher temperature (T_H) both film and substrate have length l_{TH} and, after cooling to a lower temperature (T_L), they have length l_{TC}. If they were not bonded, then, on cooling to T_L, the substrate and film would experience free thermal strains to lengths l_{subs} and l_{film} respectively:

$$\varepsilon_{thermal}^{subs} = \alpha_{subs} \Delta T \tag{8.1}$$

and

$$\varepsilon_{thermal}^{film} = \alpha_{film} \Delta T \tag{8.2}$$

where α_{subs} and α_{film} are the (assumed constant) linear thermal-expansion coefficients for the substrate and film, respectively, and $\Delta T = T_L - T_H$

and is thus negative in sign. Because they are bonded, the film and substrate also experience mechanical strains due to residual stress:

$$\varepsilon_{mech}^{subs} = \frac{\sigma_{subs}(1 - \nu_{subs})}{E_{subs}} \tag{8.3}$$

and

$$\varepsilon_{mech}^{film} = \frac{\sigma_{film}(1 - \nu_{film})}{E_{film}} \tag{8.4}$$

where Hooke's law for an equal biaxial state of stress has been used. Because the film and substrate are bonded, their total axial strains must be the same. In other words,

$$\varepsilon_{thermal}^{subs} + \varepsilon_{mechanical}^{subs} = \varepsilon_{thermal}^{film} + \varepsilon_{mechanical}^{film} \tag{8.5}$$

Performing a force balance on the specimen, where t_{subs} and t_{film} are the thickness of the substrate and that of a single film, respectively, yields

$$\sigma_{subs} t_{subs} + 2\sigma_{film} t_{film} = 0 \tag{8.6}$$

Combining these equations yields

$$\alpha_{subs} \Delta T - \frac{2\sigma_{film} t_{film}(1 - \nu_{subs})}{t_{subs} E_{subs}} = \alpha_{film} \Delta T + \frac{\sigma_{film}(1 - \nu_{film})}{E_{film}} \tag{8.7}$$

or

$$\sigma_{film} = -\frac{(\alpha_{film} - \alpha_{subs})\Delta T}{2t_{film}(1 - \nu_{subs})/(t_{subs} E_{subs}) + (1 - \nu_{film})/E_{film}} \tag{8.8}$$

If there is no Poisson-ratio mismatch between the substrate and the film (i.e. $\nu_{subs} = \nu_{film} = \nu$), the formula simplifies to

$$\sigma_{film} = -\frac{E_{film}(\alpha_{film} - \alpha_{subs})\Delta T}{(1 - \nu)(1 + 2t_{film} E_{film}/(t_{subs} E_{subs}))} \tag{8.9}$$

Note that, if the system is cooling, ΔT is negative; thus, if α_{film} is less than α_{subs}, the stress in the film will be negative (compressive), whereas if α_{film} is greater than α_{subs}, the stress will be positive (tensile). The sign of the film stress will be reversed if the specimen is heated rather than cooled. (The CTE for most materials is positive. However, some materials, e.g. specific polymers, have negative CTEs.)

For the case of a thin film on just one side of a thick substrate, it may still be reasonable to assume that there is no bending deformation of the laminate. In such cases, Equation (8.9) can be used if the factor of 2 in the last term of the denominator is removed. In cases where the film is very thin relative to the thickness of the substrate, the last term in the denominator of Equation (8.9) may be neglected, yielding

$$\sigma_{film} = -\frac{E_{film}(\alpha_{film} - \alpha_{subs})\Delta T}{1 - \nu} \tag{8.10}$$

which corresponds to the physical case of residual stresses in the film inducing essentially no deformation in the substrate.

Equation (8.9) is general and, therefore, describes also the stresses in the layers when a thin adhesive is used to join two substrates as in Figure 8.1(b). In this case the layer referred to above as the "substrate" becomes the adhesive, and the layers referred to as the "films" now become the bodies being joined. If these two bodies are identical and the adhesive layer is very thin, Equation (8.9) becomes

$$\sigma_{body} = -\frac{E_{body}(\alpha_{body} - \alpha_{adh})\Delta T}{1 - \nu} \frac{t_{adh}}{t_{body}} \tag{8.11}$$

and through the force balance, Equation (8.6), the stress in the adhesive becomes

$$\sigma_{adh} = -\frac{E_{body}(\alpha_{body} - \alpha_{adh})\Delta T}{1 - \nu} \tag{8.12}$$

8.1.3 Applied stress

Applied stresses result from external loading, e.g. the use of an adhesively bonded joint in a mechanical device that supports a load. Most of the techniques for adherence testing (tensile, indentation, etc.) involve application of an external load.

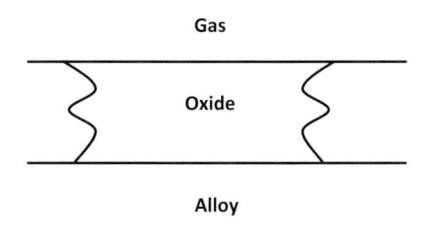

Figure 8.3 A schematic diagram of the cracking of an oxide layer that is subjected to a tensile stress.

8.2 Response to stress

The formation, thermal and applied stresses generated in a bimaterial system may be accommodated by a number of mechanisms. The most important are

(a) cracking of one of the layers,
(b) separation of the layers and
(c) plastic deformation of one or both layers.

All of these mechanisms have been observed to operate in various systems. The following discussion describes typical behavior for the case of a thin film on a thick substrate (Figure 8.1(a)) as exemplified by an oxide layer growing on a metal.

Film cracking occurs when the film is put into tension, as shown schematically in Figure 8.3. Oxides forming on the alkali metals form under tension because the molar volume of the oxide is smaller than that of the metal it replaces, i.e. the oxide–metal volume ratio is less than unity for these systems (Table 8.1).

Most often the oxides on metals and alloys are in compression because the growth stresses tend to be compressive and, in particular, the thermal stresses, when they develop on cooling, are compressive because of the sign of the thermal-expansion mismatch between the metal and oxide in Equation (8.10) (Table 8.2). Generally, the CTE of the metal is larger than that of the corresponding oxide.

Table 8.1 *Oxide–metal volume ratios of some common metals* [5]

Oxide	Oxide–metal volume ratio
K_2O	0.45
MgO	0.81
Na_2O	0.97
Al_2O_3	1.28
Cu_2O	1.64
NiO	1.65
FeO (on α-Fe)	1.68
CoO	1.86
Cr_2O_3	2.07
Fe_3O_4 (on α-Fe)	2.10
Fe_2O_3 (on α-Fe)	2.14

Table 8.2 *Linear coefficients of thermal expansion of metals and oxides and the ratio* $\alpha_{Metal}/\alpha_{Oxide}$ [5]

System	Oxide coefficient	Metal coefficient	Ratio
Fe/FeO	12.2×10^{-6}	15.3×10^{-6}	1.25
Fe/Fe$_2$O$_3$	14.9×10^{-6}	15.3×10^{-6}	1.03
Ni/NiO	17.1×10^{-6}	17.6×10^{-6}	1.03
Co/CoO	15.0×10^{-6}	14.0×10^{-6}	0.93
Cr/Cr$_2$O$_3$	7.3×10^{-6}	9.5×10^{-6}	1.30
Cu/Cu$_2$O	4.3×10^{-6}	18.6×10^{-6}	4.32
Cu/CuO	9.3×10^{-6}	18.6×10^{-6}	2.00

Figure 8.4 presents schematic diagrams of how an oxide film may respond to a compressive stress. One possibility is for the oxide and metal to separate at the interface by buckling of the oxide (Figure 8.4(a)), i.e. the oxide *spalls*. The spallation of a compressively stressed film will occur when the elastic strain energy stored in the intact film exceeds

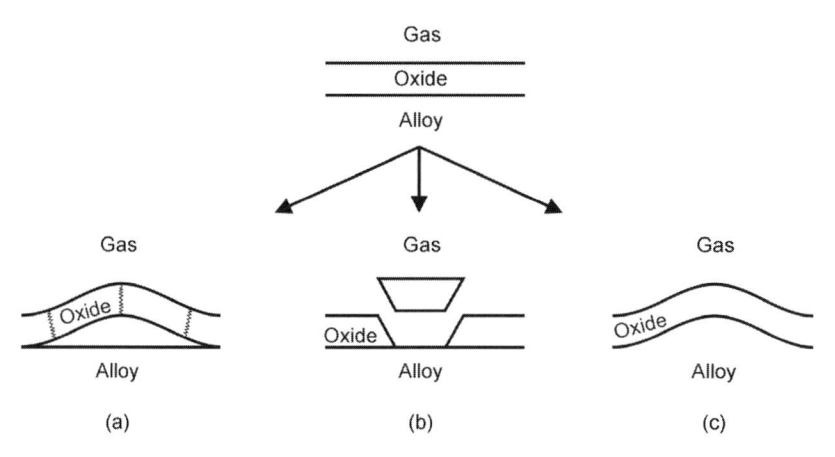

Figure 8.4 A schematic diagram showing possible responses of an oxide layer that is subjected to an in-plane compressive stress.

the fracture resistance, Γ_c, of the interface. On denoting by E and v the elastic modulus and Poisson's ratio of the film, by t the film thickness, and by σ the equal biaxial residual stress in the film, the elastic strain energy stored in the film per unit area is $(1 - v)\sigma^2 t/E$. The criterion for failure then becomes [6]

$$G = (1 - v)\sigma^2 t/E \geq \Gamma_c \qquad (8.13)$$

where G is called the strain-energy release rate. According to this criterion, which is a necessary but not sufficient condition, a film will spall if the stress is high, the film is thick, or the interfacial free energy is high (the work of adhesion is low). Decohesion of films under compression, however, requires either a buckling instability or development of a wedge crack (Figure 8.4(b)) in order to generate spallation [7]. According to elastic mechanics, buckling of a thin film under biaxial compression to form an axisymmetric buckle of radius a will occur at a critical stress, σ_c, given by

$$\sigma_c = 1.22 \frac{E}{1 - v^2} \left(\frac{t}{a}\right)^2 \qquad (8.14)$$

However, such a buckle is stable and will not propagate to cause decohesion failure by delamination unless the strain-energy release rate also satisfies Equation (8.13). Examples of buckling of alumina films from an Fe–Cr–Al alloy are presented in Chapter 5 of Reference [2].

The buckling stress, Equation (8.14), increases as the square of the film thickness such that, for thick films, buckling might not be feasible. In this case, shear cracks can form in the film and, if Equation (8.13) is satisfied, lead to scale spallation by a "wedging mechanism", which is shown schematically in Figure 8.4(b).

In some situations, if Γ_c is high and the substrate is relatively weak, the compressive stresses can be accommodated by simultaneous deformation of the film and alloy without spallation. This phenomenon is shown schematically in Figure 8.4(c).

8.2.1 The relation of the fracture energy and the work of adhesion

The *work of adhesion* was described in Chapter 3. This is the work required to separate two adherent condensed phases, as illustrated in Figure 3.4,

$$W_{ad} = \gamma_{2V} + \gamma_{1V} - \gamma_{12} \tag{8.15}$$

The work of adhesion is related to the contact angle through the Young–Dupré equation,

$$W_{ad} = \gamma_{1V}(1 + \cos\theta_Y) \tag{8.16}$$

In the ideal case the *fracture energy*, Γ, is just the work of adhesion,

$$\Gamma = W_{ad} \tag{8.17}$$

The actual amount of work that must be performed to separate two phases, however, is usually much larger than the work of adhesion because of the plastic dissipation, W_p, associated with plastic deformation of one or both of the phases which are being separated,

$$\Gamma = W_{ad} + W_p \qquad (8.18)$$

The deformation can occur by dislocation plasticity at low temperature [8] or by creep at high temperature [9].

There have been relatively few studies of the relation between fracture energy and work of adhesion. One exception is the work of Lipkin et al. [8] for interfaces between gold and sapphire (Al_2O_3). There are no reaction products when gold is bonded to sapphire, and an atomically sharp interface can be maintained between the two phases. Sandwich specimens similar to Figure 8.1(b) were prepared by diffusion bonding, with sapphire plates constituting the two substrates and gold the "adhesive". Interfacial voids were formed, presumably by entrapped air, which allowed the contact angle to be measured and the work of adhesion to be determined from Equation (8.16) using the surface energy of gold. The value determined was $W_{ad} \approx 0.6$ J/m² at the bonding temperature (1,000 °C), which was extrapolated to $W_{ad} \approx 0.9$ J/m² at room temperature. Additional sandwiches were machined to provide test specimens for the double-cleavage drilled compression (DCDC) test, for measuring the fracture energy of the Au/Al_2O_3 interface. The DCDC test configuration is shown in Figure 8.5. Numerical analysis for the geometry shown yields the strain-energy release rate as [8]

$$G = \frac{\pi \sigma_{appl}^2 R(1-\nu^2)}{E} \left[\frac{W}{R} + \left(0.235 \frac{W}{R} - 0.259 \right) \frac{a}{R} \right]^{-2} \qquad (8.19)$$

where σ_{appl} is the applied compressive stress and the other parameters are indicated in Figure 8.5. During the application of the load the propagation of the crack was observed optically through the clear sapphire plates to determine the critical energy-release rate. These room-temperature measurements yielded values of $\Gamma \approx 250$ J/m². Thus, Γ was two orders

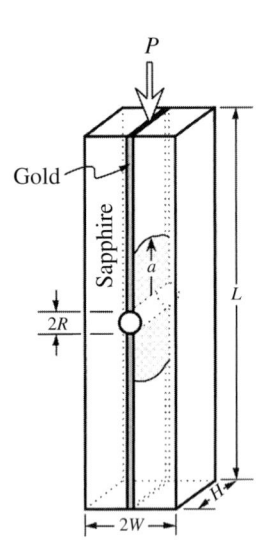

Figure 8.5 A schematic diagram of the DCDC test. (From [8], reprinted with permission of Elsevier.)

of magnitude larger than W_{ad}. This difference was ascribed to a large value of W_p resulting from dislocation plasticity of the gold.

8.2.2 The effect of adsorption on the work of adhesion and fracture energy

The experiments described above were repeated for gold/sapphire interfaces that had been contaminated with carbon by annealing the specimens in a sealed silica ampoule filled with graphite powder [8]. This treatment decreased the room-temperature work of adhesion from 0.9 to 0.6 J/m². The carbon treatment had a more dramatic effect on the fracture energy, which was reduced from 250 J/m² to 1–2 J/m². Thus the carbon heat treatment reduced the fracture energy to a value similar to the work of adhesion. Lipkin *et al.* [8] combined the above results with others from the literature to make the important point that the fracture energy scales with the work of adhesion even though the absolute value of Γ can be much larger than that of W_{ad}. This point will be revisited in the following case study of superalloys in gas turbines.

One further point must be made regarding the effect of adsorption on the work of adhesion. Lipkin *et al.* [8] and others tend to describe the reduction in W_{ad} solely in terms of adsorption to the interface. This is fundamentally incorrect in that adsorption to an interface occurs only if the adsorbing species decreases the interfacial energy (see Chapter 3). This alone would increase the work of adhesion. Therefore, there must be a greater reduction in one of the component surface energies. For example, the work of adhesion for the Au/Al$_2$O$_3$ interface is

$$W_{ad} = \gamma_{Au} + \gamma_{sapphire} - \gamma_{Au/sapphire} \qquad (8.20)$$

Thus, if carbon is adsorbed to the interface, this will reduce $\gamma_{Au/sapphire}$. If it also reduces W_{ad}, then it must be adsorbing to either the Au surface or the sapphire surface, or both, with an even larger reduction in γ_{Au} and/or $\gamma_{sapphire}$. This situation is analogous to the phenomena of grain-boundary embrittlement described in Chapter 7.

8.3 Case study – protective layers on superalloys in gas turbines

8.3.1 Formation and adhesion of protective oxide layers

One of the most demanding applications of materials is for structural components in the hot sections of gas turbines, which are used for propulsion (jet engines) and power generation. These components experience high stresses at high temperatures in oxidizing atmospheres. The materials which to date have exhibited the best combination of strength and oxidation resistance are the Ni-base superalloys, which are Ni–Al alloys to which additional strengthening elements are added. The binary Ni–Al temperature–composition diagram was presented as Figure 5.3. These multicomponent alloys contain sufficient aluminum to provide precipitates that are based on the compound Ni$_3$Al (termed γ'), which strengthen

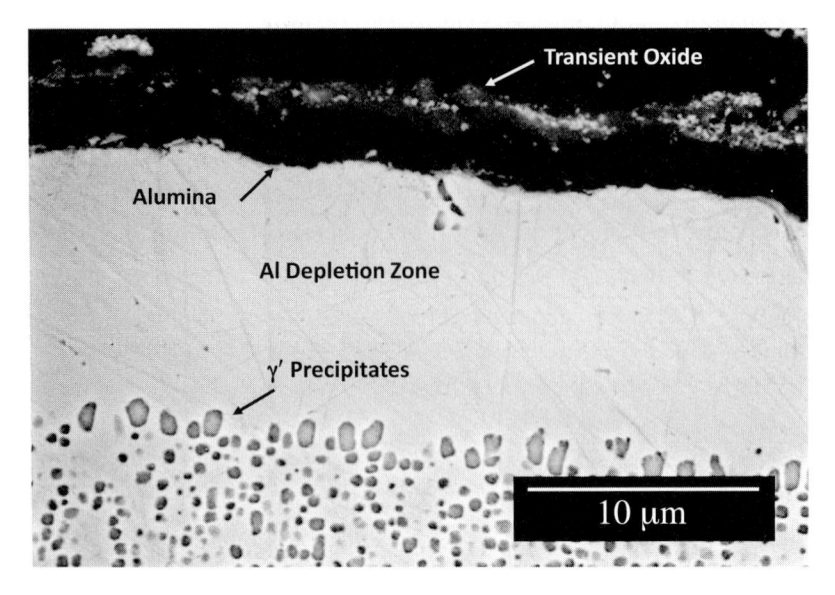

Figure 8.6 A cross-section of a nickel-base single-crystalline superalloy after high-temperature oxidation.

the alloys, and to form a surface layer of Al_2O_3, which provides protection against oxidation. The formation of alumina surface layers is described in Chapter 5 of Reference [2]. Figure 8.6 is a cross-section of a single-crystalline superalloy after high-temperature oxidation, which shows important features of these alloys. The strengthening γ' precipitates are visible at the bottom of the figure. The protective alumina layer is visible on the surface. This is similar to the schematic diagram in Figure 8.1(a). The alumina layer is covered by another layer of *transient oxide*, which had formed before the alumina became continuous. There is a zone below the alumina layer in which the γ' precipitates have dissolved. This is the result of aluminum diffusing out to grow the external alumina layer.

The adherence of the alumina film to the alloy is very important. If this layer should spall, the aluminum-depleted alloy layer will be exposed to the oxidizing atmosphere and will generally form the transient oxides,

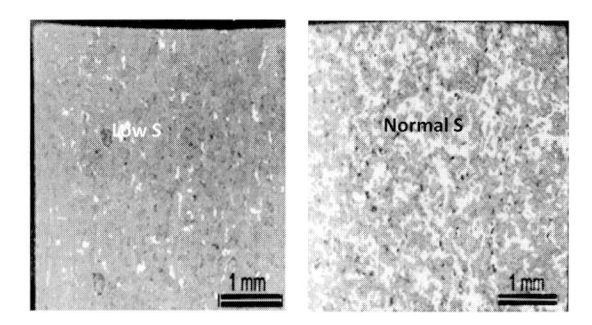

Figure 8.7 Surface micrographs of low-sulfur (<1 ppm) and normal-sulfur (≈8 ppm) superalloys after 600 one-hour cycles of oxidation at 1,100 °C.

which are not as protective as the alumina. Spalling during isothermal oxidation is not likely, since the formation (growth) stresses in the alumina layer are small (usually not more than a few hundred MPa) [9]. However, the thermal stresses resulting from cooling from the oxidation temperatures are large (often several GPa). Figure 8.7 shows the surfaces of two nearly identical superalloys after cyclic oxidation at 1,100 °C. The white areas are exposed alloy from which the alumina has spalled. The only difference between the alloys is that the one on the right has a sulfur content typical of such superalloys (≈8 ppm), whereas the one on the left has been desulfurized to levels below 1 ppm. There is much more spallation of the oxide from the alloy with the higher sulfur content. This is the result of the sulfur decreasing the fracture energy of the alumina. The discovery of this "sulfur effect" provides a good example of the practical significance of understanding surface thermodynamics. The production of low-sulfur superalloys is now a standard industrial practice.

Smialek has summarized the sulfur effect [10], which is generally explained in terms of the sulfur segregating to the alloy/oxide interface and lowering the work of adhesion. Sulfur has been shown to lower the fracture energy of artificial Ni/alumina interfaces from $\Gamma \approx 100$ J/m^2 to $\Gamma \approx 10$ J/m^2 [11]. A question remains regarding the mechanism of the

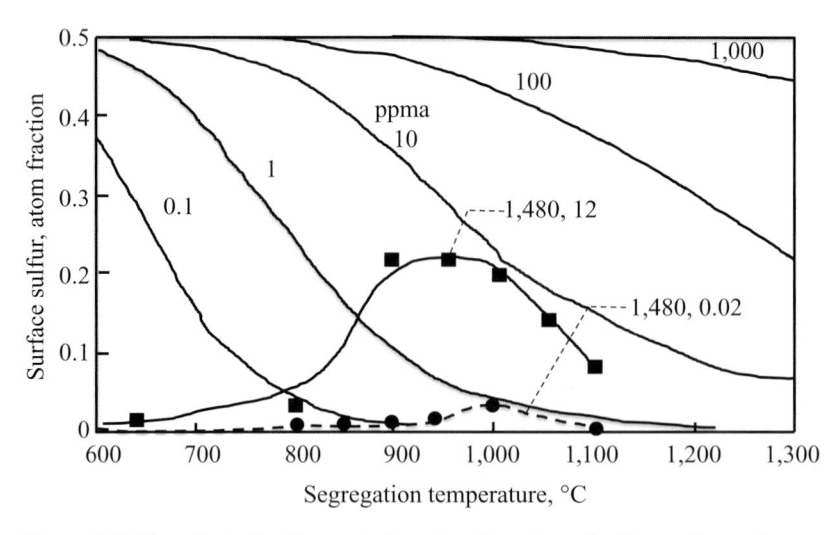

Figure 8.8 The effect of sulfur content on the adsorption of sulfur on the surface of pure Ni and a nickel-base superalloy. The family of lines corresponds to surface segregation of sulfur on Ni as a function of the sulfur concentration in the metal calculated using the Langmuir isotherm (ppma, parts per million, atomic). The lines with data points indicate experimental measurements of segregation on a nickel-base superalloy with two different sulfur contents. The square points were measured for a sulfur content of 12 ppma and the circular points were measured for a sulfur content of 0.02 ppma. (From [10], reprinted with permission of John Wiley & Sons.)

decrease in fracture energy. Sulfur has indeed been shown to segregate to interfaces between Ni and alumina [12]. However, sulfur is also known to segregate very strongly to the free surface of Ni and Ni-containing alloys. Figure 8.8 shows the surface segregation of sulfur on Ni as a function of the sulfur concentration in the metal expressed using the Langmuir isotherm and experimental measurements of segregation on a nickel-base superalloy with two different sulfur contents [10]. If sulfur is being adsorbed in the Gibbsian sense then it is lowering the interfacial energy. Therefore, if it is lowering the work of adhesion it must be lowering the surface energy of the Ni and or alumina even more. Evans has treated this aspect of the problem using finite-element analysis [13]. The results

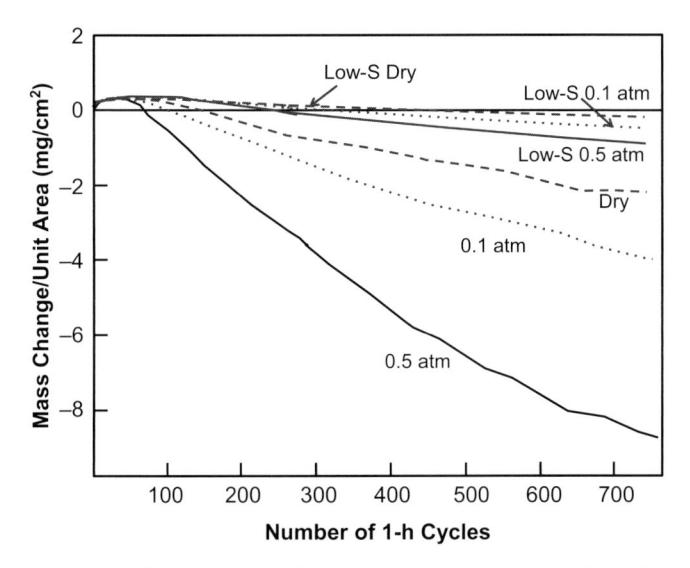

Figure 8.9 Cyclic oxidation kinetics (expressed in terms of specific mass change versus number of cycles) for a normal-S and a low-S superalloy in dry air and air containing different partial pressures of water vapor.

indicate that sulfur lowers the work of adhesion, and that the fracture energy scales with the work of adhesion and Γ is generally about a factor of 64 larger than W_{ad}.

Another interesting effect is the influence of water vapor on the spallation of the alumina. Figure 8.9 shows the cyclic oxidation kinetics of the two superalloys in different atmospheres as a plot of mass gain per unit surface area versus number of cycles. If there is no oxide spallation, the mass will increase because of the oxygen being incorporated into the oxide. However, if there is spallation, the mass change will be the composite of the gain associated with oxygen uptake and loss from the oxide spalling from the surface. As spalling becomes significant, the net mass change becomes negative. Figure 8.9 shows that the mass change for a low-sulfur alloy with the very adherent alumina film is largely unaffected by the water vapor. However, the mass losses (oxide spallation)

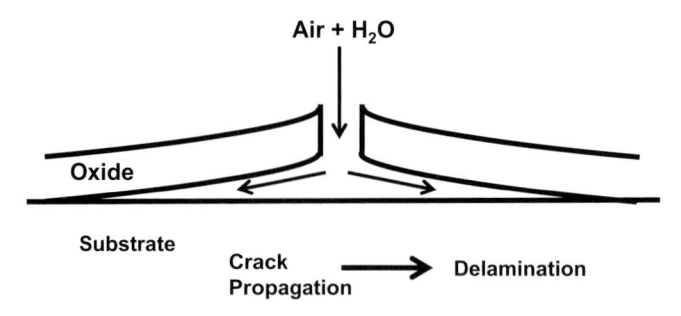

Figure 8.10 A schematic diagram showing the effect of water vapor on the delamination of an oxide from a metal.

for the normal-sulfur alloy become more significant the higher the partial pressure of the water vapor. This effect has been explained in terms of a stress-corrosion cracking phenomenon similar to that for slow crack growth in monolithic ceramics described in Chapter 7 [14, 15] as illustrated in Figure 8.10. When the adherence of alumina is poor in dry air (e.g. for the normal-S alloy) the separation of bonds at the crack tip is greater and the adsorption of water molecules can accelerate crack growth. The tighter binding at the oxide/alloy interface for the low-S alloy minimizes this effect.

8.3.2 Multilayer systems – thermal barrier coatings

Thermal barrier coatings (TBCs) are ceramic coatings that are applied to components for the purpose of insulation rather than oxidation protection. See, for example, Chapter 10 in Reference [2]. The use of an insulating coating, coupled with internal air-cooling of the component, lowers the surface temperature of the component with a corresponding decrease in the creep and oxidation rates of the component. The use of TBCs has resulted in a significant improvement in the efficiency of gas turbines. Typical systems, shown schematically in Figure 8.11, consist of a nickel-base superalloy substrate coated with MCrAlY (M = Ni, Co) or

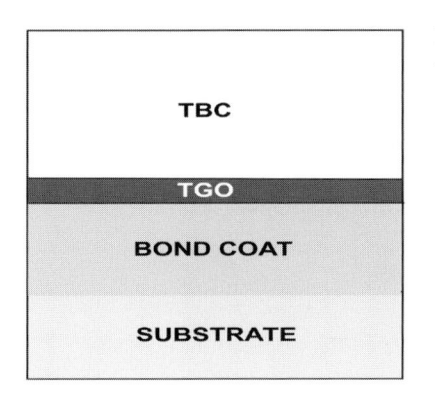

Figure 8.11 A schematic drawing of a thermal-barrier-coating system.

a diffusion aluminide bond coat that forms an alumina layer (thermally grown oxide, TGO) onto which is deposited an yttria-stabilized zirconia (YSZ) TBC. The TBC can be deposited by air plasma spraying (APS), or electron beam physical vapor deposition (EBPVD). The EBPVD coatings are used for the most demanding applications, such as the leading edges of airfoils. Figure 8.12 presents a cross-section micrograph of a typical EBPVD TBC and Figure 8.13 shows the cross-section of an APS TBC. The EBPVD coating consists of columnar grains, which are separated by channels. These channels are responsible for the EBPVD TBCs having a high strain tolerance. The APS coating consists of layers of *splats* with clearly visible porosity and is microcracked. This microcracking is necessary for strain tolerance.

The major challenge for the application of TBCs is coating durability, particularly the resistance of the coating to spalling [16].

As can be seen from Figure 8.11, a TBC system is a multilayer system with each layer having a different CTE. Typical values are as follows:

YSZ top coat,	11×10^{-6} K^{-1}
Alumina TGO,	8.8×10^{-6} K^{-1}
Bond coat,	14×10^{-6} K^{-1}
Superalloy substrate,	15×10^{-6} K^{-1}

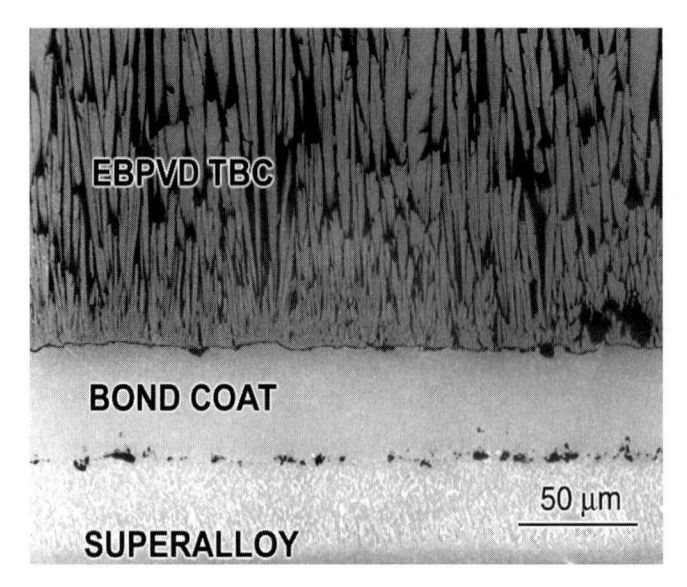

Figure 8.12 A cross-section of a thermal barrier coating fabricated by electron-beam physical vapor deposition (EBPVD).

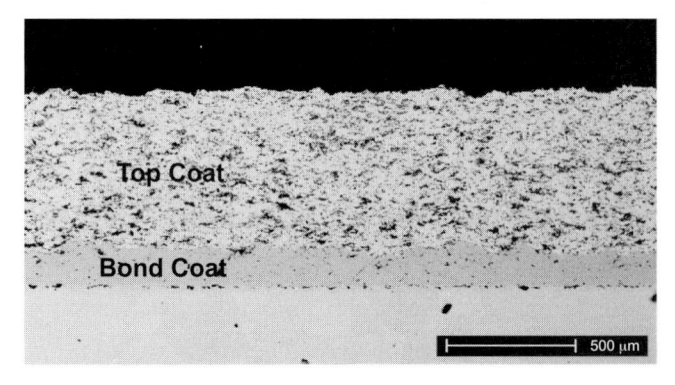

Figure 8.13 A cross-section of a thermal barrier coating fabricated by air plasma spraying (APS).

Since the superalloy substrate, which is generally much thicker than the other layers, has a higher CTE, the outer layers are loaded in compression when cooled from a high temperature. This can result in failures similar to those shown in Figure 8.4 for a bilayer system. Often the failure

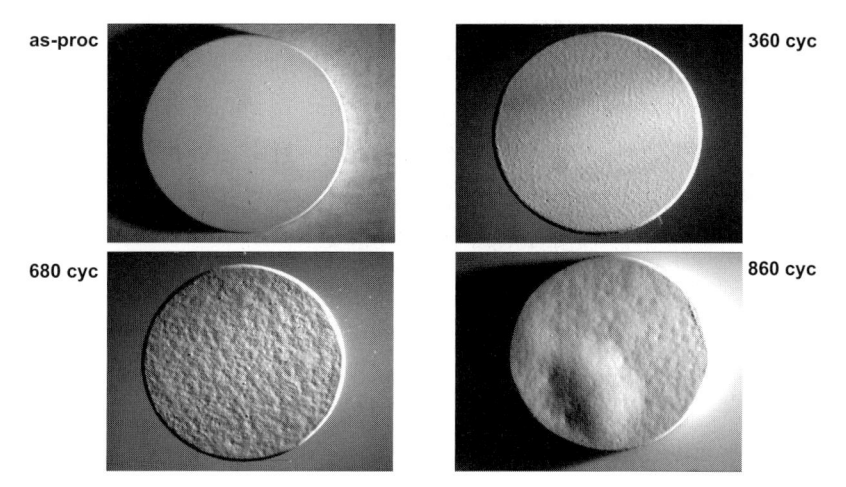

Figure 8.14 Surface photographs showing the development of a buckle on a TBC system: as-proc, as-processed, cyc, cycles.

occurs by buckling. Figure 8.14 shows optical images of an EBPVD TBC deposited on a platinum-modified aluminide bond coat, which was exposed using one-hour thermal cycles between 1,100 °C and room temperature. The surface is smooth prior to exposure, but small buckles are observed after 360 cycles. These small buckles continue to coalesce with continued cycling into a larger buckle after 860 cycles. When this buckle reaches a critical size the coating will spall. Figure 8.15 shows the surface of a separate specimen on which the coating has failed by buckling. Figure 8.16 shows the specimen after removal of the coating. The exposed surface is that of the bond coat, which indicates that the separation propagated between the TGO and the bond coat.

In order to quantify the fracture energy for such a coating, a test that causes the coating to fail by the same mechanism is required. This can be accomplished using an indentation test, which was first developed for bilayer systems [17]. The test consists of indenting the TBC system with a Brale-type conical indenter (a Rockwell hardness tester or mechanical testing machine can be used for this purpose) [18, 19].

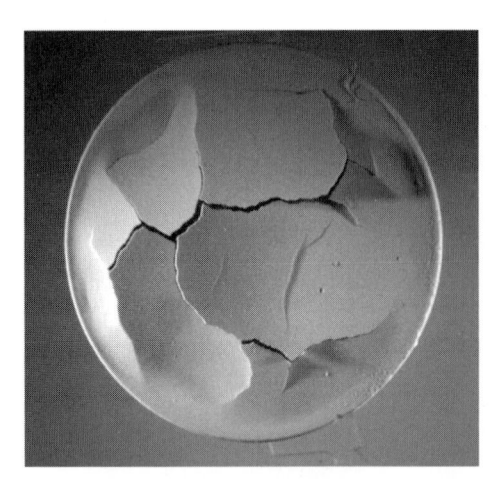

Figure 8.15 A surface photograph showing the failure of a buckled TBC.

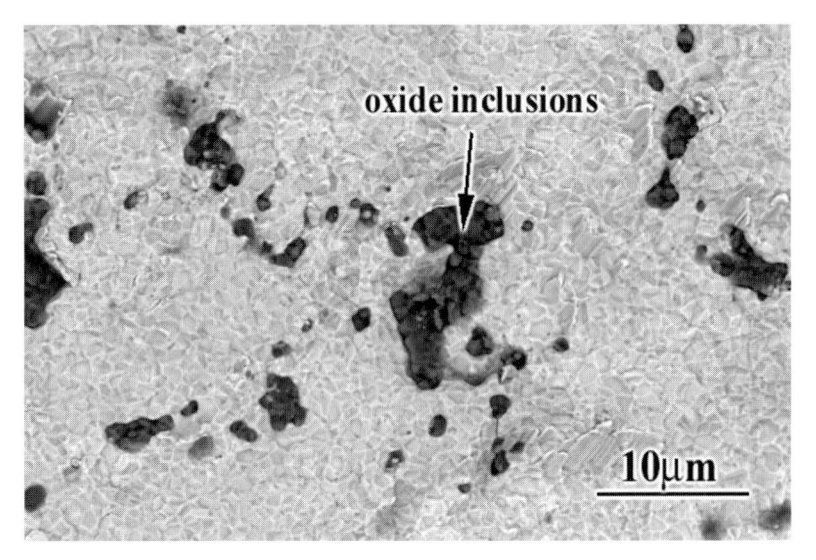

Figure 8.16 A micrograph showing the surface from which a TBC has spalled.

The test is shown schematically in Figure 8.17. The indenter penetrates the TBC and oxide layers, and plastically deforms the metallic bond coat and substrate below. As illustrated in Figure 8.18 (viewed from above), this induces an axisymmetric debonding of the TBC and oxide layers similar to that for a thermally induced failure by buckling.

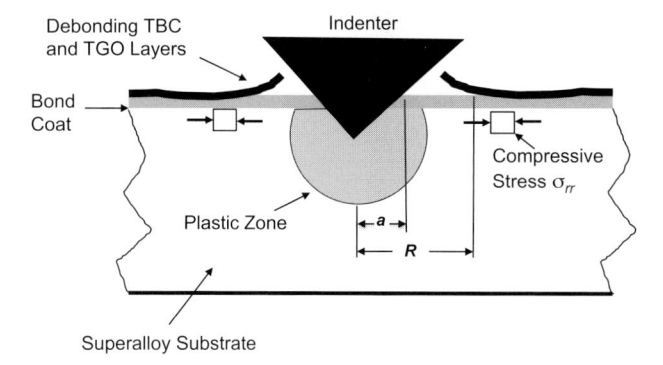

Figure 8.17 A schematic diagram of the indentation test for measuring TBC adherence. Here a is the contact radius and R is the debonding radius.

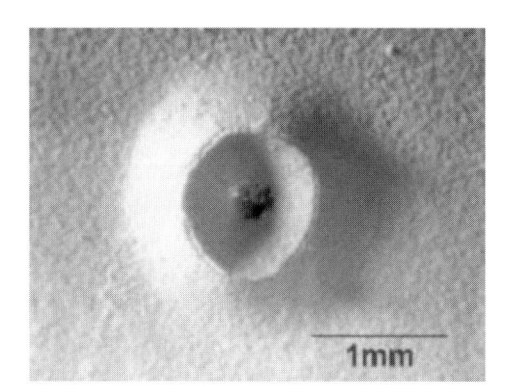

Figure 8.18 The surface of a TBC following indentation.

Systems with poor adhesion between the oxide and the bond coat show large debonding radii. Systems with good adhesion will show small debonding radii or no debonding at all. Fracture-mechanics analyses of the indentation test have been performed to allow determination of an interfacial toughness (a value of the critical stress-intensity factor, K_c, or critical energy-release rate, G_c) for the interface between the oxide and bond-coat layers, for a measured debonding radius [18–20]. Accurate measurement of the debonding radius is required. This may be done by placing the specimen, without a conductive coating, in a scanning

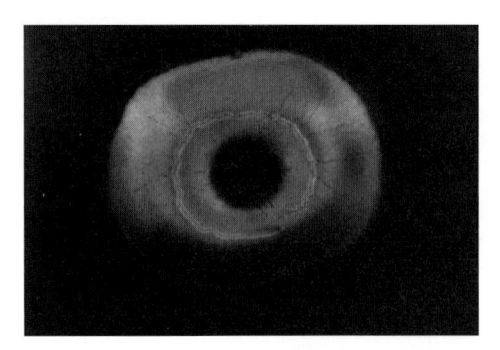

Figure 8.19 A charging image of the specimen in Figure 8.18.

electron microscope. The region of the coating which is no longer in contact with the underlying metallic layers will not be able to efficiently conduct away the bombarding electrons and hence will "charge", which allows the debonding to be imaged, Figure 8.19.

This indentation test allows tracking of toughness changes in TBC systems as a function of thermal exposure, including the ability to perform many indentations on a single specimen, as its exposure times are increased. For example, Figure 8.20 shows a plot of "apparent" losses of interfacial toughness for EBPVD TBC/PtAl bond-coat systems exposed to 1,200 °C, 1,135 °C and 1,100 °C isothermal exposures in laboratory air, as determined by room-temperature indentation tests. (Note that there is some variability in the toughness for as-processed specimens.) The times of 60, 500 and 1,000 hr are approximate times to spontaneous spallation at temperatures of 1,200 °C, 1,135 °C and 1,100 °C, respectively. At such times, the interfacial fracture toughness matches the applied stress-intensity factor due to thermal strains alone, which is approximately equal to 1 MPa m$^{1/2}$. These data indicate that substantial toughness loss occurs at a fraction of the time needed for spontaneous failure to occur. The conversion between K (in units of MPa m$^{1/2}$) and G (in units of J/m^2) is [21]

$$K = \sqrt{\frac{G}{3.58}}$$

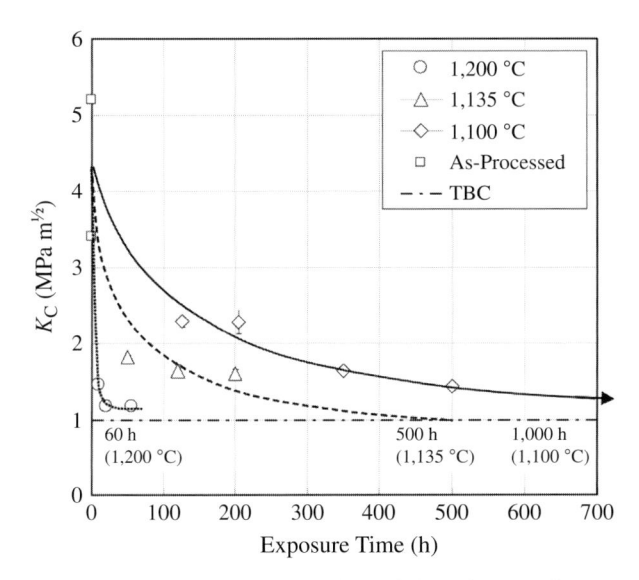

Figure 8.20 A plot showing the degradation of TBC adherence with exposure time at three temperatures.

Therefore, Figure 8.20 indicates that the fracture energy for an as-processed coating is $\Gamma \approx 60$ J/m². This value is of similar magnitude to those measured for Ni/alumina interfaces in hot-pressed specimens [12], i.e. it is much larger than the work of adhesion. This fracture energy decreases with exposure to values corresponding to $\Gamma \approx 4$ J/m², i.e. Γ is approaching W_{ad}. This means that the TBC can undergo essentially ideal fracture and will be more prone to failure.

8.4 Summary

In this chapter the relation of the fracture energy (Γ) to the work of adhesion (W_{ad}) in bilayer and multilayer systems has been described. Generally Γ is significantly larger than W_{ad} but tends to scale with W_{ad}. The effects of adsorption on film adherence have been described, and a case study of the relevance of film adherence to advanced superalloys and thermal barrier coatings has been presented.

8.5 References

[1] D. A. Jones, *Principles and Prevention of Corrosion* (New York: Macmillan Publishing Co., 1991).
[2] N. Birks, G. H. Meier and F. S. Pettit, *Introduction to High Temperature Oxidation of Metals*, 2nd edn. (Cambridge: Cambridge University Press, 2006).
[3] S. Ebnesajjad, *Surface Treatment for Adhesion Bonding* (Norwich, NY: William Andrew Publishing, 2006).
[4] S. P. Timoshenko, Analysis of bimetal thermostats, *J. Opt. Soc. Amer.*, **11** (1925), 233–256.
[5] P. Hancock and R. C. Hurst, The mechanical properties and breakdown of surface oxide films at elevated temperatures, in *Advances in Corrosion Science and Technology*, eds. R. W. Staehle and M. G. Fontana (New York, Plenum Press, 1974), pp. 1–84.
[6] H. E. Evans and R. C. Lobb, Conditions for the initiation of oxide-scale cracking and spallation, *Corros. Sci.*, **24** (1984), 209–222.
[7] H. E. Evans, Stress effects in high temperature oxidation of metals, *Int. Mater. Rev.*, **40** (1995), 1–40.
[8] D. M. Lipkin, D. R. Clarke and A. G. Evans, Effect of interfacial carbon on adhesion and toughness of gold-sapphire interfaces, *Acta Mater.*, **46** (1998), 4835–4850.
[9] E. Schumann, C. Sarioglu, J. R. Blachere, F. S. Pettit and G. H. Meier, High-temperature stress measurements during the oxidation of NiAl, *Oxid. Metals*, **53** (2000), 259–272.
[10] J. L. Smialek, Advances in the oxidation resistance of high-temperature turbine materials, *Surf. Interface Anal.*, **31** (2001), 582–592.
[11] J. D. Kiely, T. Yeh and D. A. Bonnell, Evidence for the segregation of sulfur to Ni–alumina interfaces, *Surf. Sci. Lett.*, **393** (1997), L126–L130.
[12] A. G. Evans, J. W. Hutchinson and Y. Wei, Interface adhesion: effects of plasticity and segregation, *Acta Mater.*, **47** (1999), 4093–4113.

[13] H. E. Evans, Predicting oxide spallation from sulphur-contaminated oxide/metal interfaces, *Oxidation Metals*, **79** (2013), 3–14.

[14] R. Janakiraman, G. H. Meier and F. S. Pettit, The effect of water vapor on the oxidation of alloys that develop alumina scales for protection, *Metall. Mater. Trans. A*, **30** (1999), 2905–2913.

[15] M. C. Maris-Sida, G. H. Meier and F. S. Pettit, Some water vapor effects during the oxidation of alloys that are α-Al_2O_3 formers, *Metall. Mater. Trans. A*, **34** (2003), 2609–2619.

[16] A. G. Evans, D. R. Mumm, J. W. Hutchinson, G. H. Meier and F. S. Pettit, Mechanisms controlling the durability of thermal barrier coatings, *Progr. Mater. Sci.*, **46** (2001), 505–553.

[17] M. D. Drory and J. W. Hutchinson, Measurement of the adhesion of a brittle film on a ductile substrate by indentation, *Proc. R. Soc. Lond.*, **452** (1996), 2319–2341.

[18] R. A. Handoko, J. L. Beuth, G. H. Meier, F. S. Pettit and M. J. Stiger, Mechanisms for interfacial toughness loss in thermal barrier coating systems, in *Durable Surfaces, Proceedings of the Materials Division Symposium on Durable Surfaces, 2000 ASME International Mechanical Engineering Congress and Exposition*, Orlando, November, eds. D. R. Mumm, M. Walter, O. Popoola and W. O. Soboyejo (Zurich: Trans Tech Publications, 2000), pp. 165–183.

[19] A. Vasinonta and J. L. Beuth, Measurement of interfacial toughness in thermal barrier coating systems by indentation, *Eng. Fracture Mech.*, **68** (2001), 843–860.

[20] M. J. Stiger, G. H. Meier, F. S. Pettit, Q. Ma, J. L. Beuth and M. J. Lance, Accelerated cyclic oxidation testing protocols for thermal barrier coatings and alumina-forming alloys and coatings, *Mater. Corrosion*, **57** (2006), 1–13.

[21] Q. Ma, Indentation methods for adhesion measurement in thermal barrier coating systems, Ph.D. Thesis, Carnegie Mellon University, 2004.

8.6 Study problems

1. A thick piece of an Fe–18 wt% Cr–6 wt% Al alloy is oxidized at 1,100 °C until an α-alumina layer 1 μm thick has formed.

Neglecting any growth stresses and any stress relaxation, what is the sign and magnitude of the stress in the alumina when the specimen has been cooled to room temperature?

Use the following data:

$\alpha(\text{FeCrAl}) = 15 \times 10^{-6} \text{ K}^{-1}$

$\alpha(\text{Al}_2\text{O}_3) = 8.8 \times 10^{-6} \text{ K}^{-1}$

$E(\text{Al}_2\text{O}_3) = 400$ GPa at 298 K

$v(\text{Al}_2\text{O}_3) = 0.24$

2. Consider an APS thermal barrier coating similar to that in Figure 8.13, which has been exposed in an oxidizing environment at 1,100 °C until an alumina TGO grew to a thickness of 10 μm and was then cooled to room temperature. Given the data below, calculate the elastic strain energy stored in the TGO and the YSZ topcoat:

$\alpha(\text{superalloy}) = 15 \times 10^{-6} \text{ K}^{-1}$, $t_{\text{Superalloy}} = 2$ mm

$\alpha(\text{bond coat}) = 14 \times 10^{-6} \text{ K}^{-1}$, $t_{\text{Bond coat}} = 100$ μm

$\alpha(\text{YSZ}) = 11 \times 10^{-6} \text{ K}^{-1}$, $t_{\text{YSZ}} = 300$ μm

$\alpha(\text{Al}_2\text{O}_3) = 8.8 \times 10^{-6} \text{ K}^{-1}$

$E(\text{Al}_2\text{O}_3) = 400$ GPa at 298 K

$E(\text{YSZ}) = 100$ GPa at 298 K

$v(\text{Al}_2\text{O}_3) = 0.24$

$v(\text{YSZ}) = 0.24$

Index

activity, 11, 12, 23, 24, 31, 36, 37, 171, 173, 174, 175, 176, 189
activity coefficient, 12
adhesion, i, xi, xiv, 77, 78, 210, 211, 217, 218, 219, 220, 221, 223, 231, 233, 234, 235
adsorption, i, x, xi, xiv, 42, 57, 60, 184, 186, 188, 190, 191, 192, 193, 194, 195, 196, 198, 201, 203, 204, 205, 207, 208, 211, 220, 221, 224, 226, 233

buckling, 216, 217, 218, 229, 230

capillary-rise technique, 78, 80, 151
Cassie's law, 85
chemical potential, 9, 11, 21, 153, 154, 155, 157, 171, 173, 174, 176, 187, 188, 189, 197, 204, 205
chemical reactions, 30
coarsening, x, 131, 148, 176, 177, 180
coherent interfaces, ix, 128
coincident-site lattice, 110
common-tangent construction, 22, 173
contact angle, 76, 78, 81, 82, 83, 85, 86, 87, 88, 89, 92, 93, 127, 151, 218, 219
crack propagation, 196
crack velocity, 196, 197, 198

desorption, 42, 184

enthalpy, 5, 7, 14, 35, 36, 38, 50, 72
entropy, 6, 7, 10, 16, 35, 36, 38, 43, 49, 56, 58, 67, 71, 72, 98, 140, 181, 187
equilibrium constant, 31, 193, 194
equilibrium shape, 98, 99, 101, 116

facets, 116, 117, 120
first law of thermodynamics, 3
fracture energy, xi, 211, 218, 219, 220, 223, 229, 233
fugacity, 11

Gibbs dividing surface, 40, 41, 69, 184, 187, 188
Gibbs free energy, 8, 9, 11, 16, 21, 28, 30, 32, 34, 35, 49, 50, 51, 71
Gibbs phase rule, 20, 21, 24, 25, 37
Gibbs–Duhem relation, 2
Gibbs–Thomson equation, 148, 154, 155, 156, 180
grain-boundary grooving technique, 98, 123
grain-boundary segregation, x, 195, 198, 201, 202, 207, 208
grain growth, 148, 156, 158
green steels, 206

heat, 4, 5, 7, 14, 35, 36, 195, 206, 210, 220
Helmholtz free energy, 8, 47, 48, 49, 159
Henry's law, 13
heterogeneous nucleation, 139, 166
heterogeneous system, 17
high-angle grain boundaries, 102, 108
homogeneous system, 17

ideal solution, 12, 13, 16, 28, 37, 173
ideal work of separation, 201
incoherent interfaces, ix, 133
indentation test, 229
internal energy, 4, 13, 14, 15, 18, 49, 160, 187
interphase interfaces, ix, 128

Kelvin equation, 68, 164, 179

Laplace equation, ix, 60, 69, 148, 150, 151, 152, 154, 159, 161, 162, 164, 179
line tension, 80

237

Maxwell reciprocal relations, 9, 10, 49, 71
metastability, 28
method of undetermined multipliers, 100
miscibility gap, 26, 27, 37
mixing quantity, 11
molar quantity, 2

nanoparticles, 175
non-ideal solutions, 12
non-PV work, 4, 50
non-reactive wetting, 73
nucleation, x, 139, 148, 165, 166, 167, 168, 182
number of components, 17
number of degrees of freedom, 18

partial molar quantities, 2
pendant drop technique, 152
phase, 17
phase diagram, 17
properties, 1

quasichemical solution model, 17

Raoult's law, 12
reactive wetting, 73
regular solution, 13, 15, 16, 17, 37, 66

second law of thermodynamics, 5
semicoherent interfaces, ix, 132
sessile drop, 74
solution thermodynamics, 10
spallation, 216, 218, 223, 225, 232, 234, 235
state functions, 1
strain energy release rate, 217
subgrain boundaries, 103
sulfur effect, 223

surface energy, ix, xiv, 40, 44, 45, 48, 49, 51, 52, 53, 55, 56, 57, 58, 59, 60, 63, 64, 65, 66, 68, 69, 78, 79, 93, 94, 95, 96, 99, 101, 116, 118, 119, 120, 121, 126, 127, 131, 138, 146, 148, 151, 155, 165, 166, 167, 181, 182, 188, 189, 191, 192, 201, 209, 219, 224
surface excesses, 42, 194, 201, 207
surface stress, xiv, 40, 43, 45, 47, 50, 51, 52, 53, 60, 63, 64, 65, 69, 71, 76, 94, 148, 155, 163, 165, 179, 180
surface tension, 44
surface work, 4

thermal barrier coatings, 226
thermal stresses, 212, 215, 223
third law, 6
tilt boundaries, 103, 105, 106
transient oxide, 222
triple point, 19, 64, 68, 82, 86, 92, 113, 115, 134
twin boundaries, 94, 102, 113
twist boundaries, 103, 106, 108, 109, 110, 111

van't Hoff isotherm, 31

wedge crack, 217
Wenzel equation, 81
wetting, i, viii, xiv, 68, 73, 74, 78, 80, 81, 83, 84, 87, 88, 89, 91, 114, 168
wetting angle, 68, 74
work, xi, 4, 45, 47, 50, 52, 69, 77, 78, 79, 94, 149, 201, 210, 211, 217, 218, 219, 220, 221, 223, 233
work of adhesion, viii, 77
Wulff plot, 96, 99, 101, 116, 118, 119

Young's equation, 74
Young–Dupré equation, 78

zero-creep technique, 58